The Bomb Shelter Builders Book

Floyd Delrose

Disclaimer

This publication is for entertainment purposes only. While the author has used his best efforts in preparing the book, no representation or warranty is made with respect to the accuracy or completeness of the contents within, therefore, the author is not liable for any losses, injuries, or damages. The plans and strategies contained within this book may not be suitable for every situation. Consult with a professional where appropriate.

This book is based on material from the following U.S. Government publications:

MP-15 The Family Fallout Shelter, The Office of Civil and Defense Mobilization 1959
H-12-1 Belowground Home Fallout Shelter, Federal Emergency Management Agency 1975-80
H-12-3 Home Blast Shelter, Federal Emergency Management Agency 1981-83
H-12-4 Home Shelter: Plans for a home shelter of masonry block that provides protection from nuclear fallout radiation and tornadoes, Federal Emergency Management Agency 1987

The Bomb Shelter Builders Book

Copyright © 2012 Floyd Delrose

All Rights Reserved

Contact: floyddelrose@gmail.com

Contents

Plans and Construction Overview 7

Standards in Subterranean Bomb Shelter Design 8

MP-15 Concrete Underground Bomb Shelter Plans 10

Aspects of H-12-1 Home Shelter Plans Booklet 23

H-12-4 Masonry Block Underground Bomb Shelter Overview 26

H-12-4 Materials List 31

H-12-4 Masonry Block Underground Bomb Shelter Plans 32

Sump Design and Installation 50

Foundation Drainage and Weatherproofing 53

MP-15 Aboveground Masonry Block Bomb Shelter Plans 54

Preface

The concept of burrowing underground to seek shelter or concealment is nothing new, it's been happening for millions of years in the natural world, but modern humans for the most part did not discover a need for such endeavors until the advent of high explosive bombardment introduced sometime during the past two centuries of frequent wars.

Wartime underground fortifications made of concrete became prevalent in the early 1900's many of which remain standing today long after timber framed trenches and bunkers have decomposed back into the earth.

The public interest in underground shelter construction was spurred by bombing raids that often targeted civilian areas during World War Two. This interest later reached a fever pitch during the 1950's and 60's cold war era.

Thermonuclear detonation during the Pacific tests in 1958, Hardtack-I Operation.
Photo courtesy of National Nuclear Security Administration / Nevada Site Office

In the united states, municipalities under the direction of the Federal Civil Defense Authority designated thousands of public buildings as fallout shelters but many citizens feared they would not have time to reach a local shelter before the blast occurred. So, a market for backyard bomb shelter plans and construction emerged. Between fathers office building with a fallout shelter in the city, the children's school with a fallout shelter in basement, and mom at home with a fallout shelter under the backyard, all bases were covered providing complete fallout shelter coverage.

The plans in this book are based on some of the fallout shelter designs provided to the public by government during the cold war era. Additional information on building with concrete not provided in original government materials is also included. This book will strive to produce a more complete guide for constructing the shelters than the original plans did.

Today most people hold a diminished fear of nuclear attack. This is odd when you consider how many nuclear warheads remain armed and active. Regardless of that fact, an underground shelter continues to offer the prepared individual an asset that most people will not have during a catastrophic act of nature or complete breakdown of society. It is for these reasons alone that a personal bomb shelter could be considered, the idea studied, and the investment in material and work undertaken.

Excavating a large amount of earth, building forms, and pouring concrete is a lot of work and expense, but when you consider the fact that a structure such as this will likely survive hundreds if not thousands of years after you are gone ... the work and expense suddenly seem insignificant. Perhaps even warranted.

Plans and Construction Overview

In their time, fallout shelters were offered as a method of protection from nuclear fallout (the radioactive dust that gets kicked up into the atmosphere during a nuclear explosion and spread by winds to slowly fall back to earth over the course of days to weeks after a detonation), but were also, and can also, be used for more than that. As the specter of an all out nuclear exchange with the then Soviet Russia waned, private shelters undoubtedly took on other uses. Storage, wine cellars, or early versions of the "man-cave" would probably be some common uses.

The cold war era in which these ideas blossomed is over but another cold war could happen. As long as there are feuding nations it can happen. The only way to remove the possibility of nuclear weapon use, either as sanctioned by a government or a rogue group ... would be to invent a time machine and change history. The genie is out of the bottle as they say and nuclear capable nations will continue to have the ability to manufacture mass devices of death undetected while circumnavigating international law and treatise.

Smaller more easily concealed nuclear devices that would harm by spread of fallout rather than sheer explosive force actually pose the most realistic threat in today's world. This is not to say that a nation or independent group could not fabricate and deliver a sizable nuclear bomb ... just that the odds of it happening are much less. Anyone in the blast zone of a nuclear bomb would not survive anyways, except for those rare instances when sheltered in a deep government or multi million dollar private underground complex.

With that being said, there are many other unfortunate situations where a personal underground shelter could save your life (such as tornados), but the majority of today's enthusiasts are likely studying the possibility of a shelter in preparation for some unknown threat. It's difficult to say what this threat might be. Many fear a collapse of society that arrives in the form of mass anarchy followed by government oppression or vice versa. In a scenario like this an underground shelter becomes about keeping it's inhabitants undetected from marauding bands of armed criminals, or local, state and federally appointed gunmen.

The logical downside is once an underground shelter location is discovered, flushing the inhabitants out by blocking air vents or flooding with water is a relatively simple procedure. The solution to this is easy... build your shelter in a wooded area and camouflage the entry and vent tubes.

The logistics of successfully operating a functioning underground shelter will be effected by decisions made prior to it actually being built, as is illustrated in the previous example in regard to choice of building site. It is more difficult, and requires more creativity, to conceal the entry and air vents of a personal underground shelter within the landscape of an urban or suburban area. The best way is to cover the entry and vents with something that will not attract the interest of thief's or officials. Perhaps a pile of wood chips, or some dense shrubbery.

Of course, any neighbors who watched your shelter be built will be aware of it's existence. This presents an entire laundry list of problems unto itself... as you might imagine. Relationships with neighbors usually fall into one of three categories: good, neutral, or bad. Even a friendly neighbor can turn on you in a state of panic during an emergency situation. So, unless a neighbor is sharing the financing, planning, and ultimate use of the shelter with you, I would consider all neighbors with any knowledge of your structure into a category of potential threat. The threat posed by these neighbors is not direct but rather indirect by way of who they share their knowledge with.

So, as mentioned before, the best way to avoid these headaches is to build in a rural area on a few acres of land with an access drive sufficiently cleared and prepared affording heavy construction vehicles trouble-free access.

Standards in Subterranean Bomb Shelter Design

Most of the classic cold war underground bomb shelter plans were designed for use by members of the general public operating with limited funds. It is for this reason that the designs were likely kept small and spartan in nature, just large enough for the typical family unit.

Such personal bomb shelters were usually 80-120 square foot rectangle concrete vaults with 2-4 feet fill over a slab roof. Essentially a self contained basement with 8-12 inch thick walls, a slab floor, and the aforementioned slab roof ... all steel reinforced.

These shelters are simple one room designs with the main variable being the entrance. Staircases of traditional dimension would require a substantial amount of material for a structure not much larger than the stairway itself... but a ships ladder (50-70 degree incline steps) or a fully vertical wall mounted ladder offers the builder two effective and space saving options. Both ladder types are permanently anchored as shown in the original plans but could be custom fabricated so that the lower rungs are removable and storable when not in use.

The second style of entry way is a permanent vestibule built into the side of the shelter that houses a steep staircase terminating above ground with a hatch or door. This is a better choice for use by those with reduced mobility who would not be able to descend on a ladder. Stairs are also a more convenient feature for those who plan on using their underground room on a frequent basis, carrying supplies up and down often, and so forth.

Underground Concrete and Masonry Block Built Fallout Shelter Plans Provided to Citizens by the U.S. Government 1959-87

Office of Civil and Defense Mobilization Shelter Plans

MP-15 The Family Fallout Shelter (June 1959)

MP-18 Clay Masonry Family Fallout Shelters (February 1960)

Office of Civil Defense Shelter Plans

H-7   Family Shelter Designs (January 1962)

H-12-1 Belowground Home Fallout Shelter (November 1980)

H-12-3 Home Blast Shelter (November 1983)

H-12-4 Home Shelter: Plans for a home shelter of masonry block that provides protection from nuclear fallout radiation and tornadoes (October 1987)

Note: most plans in the MP and H series of booklets illustrated lower cost expedited fallout shelters designed to be built within, or adjacent to, the basements of pre-existing structures as seen in the bottom photo on the opposite page. This is understandable as the majority of citizens had neither the funds nor inclination to take-on a full underground shelter building project. The booklets listed above included at least one plan for a fully self contained concrete or masonry block underground shelter, along with the aforementioned expedited fallout shelters.

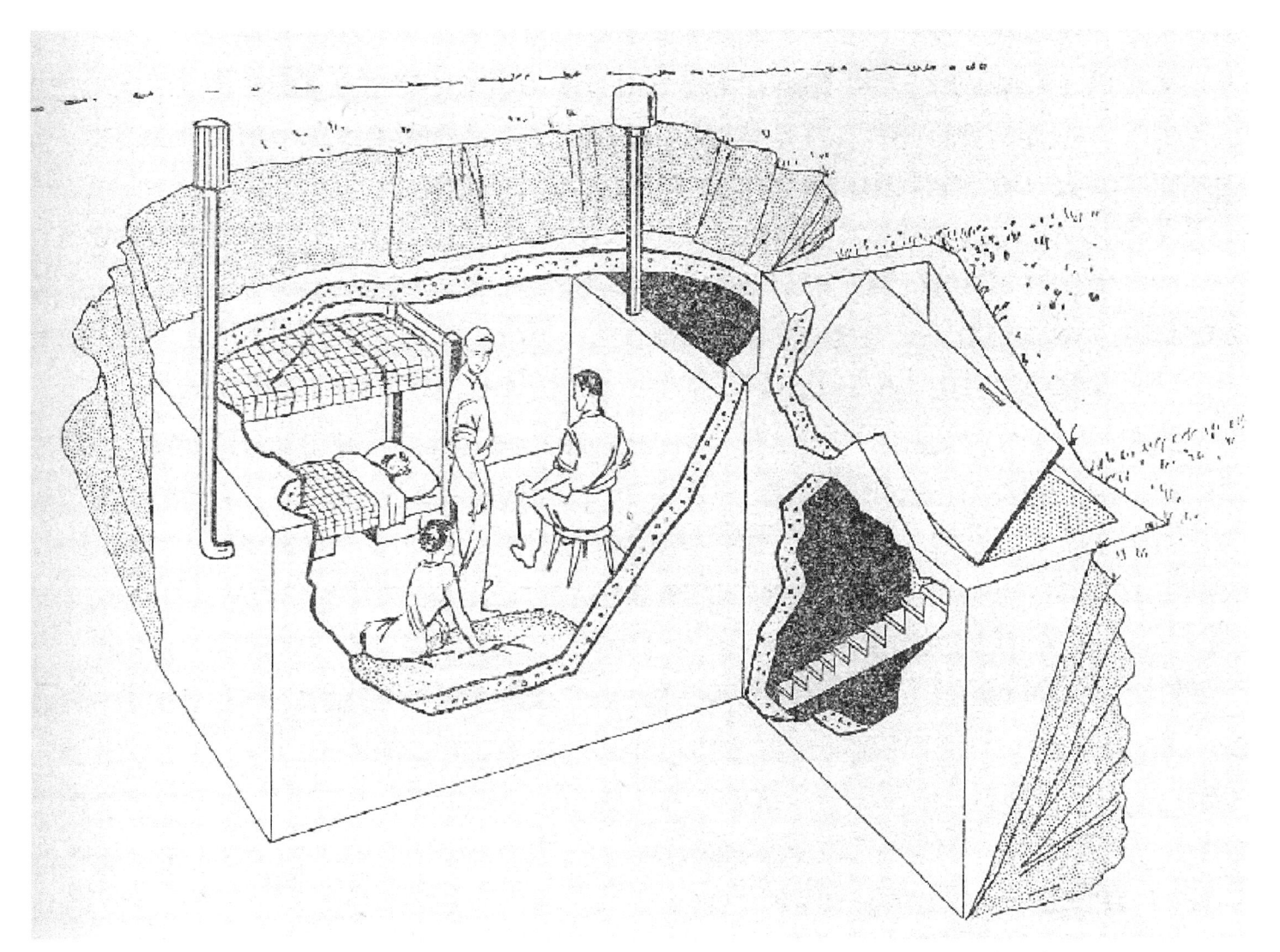

The stand-alone underground shelter plans featured in this publication are derived from Civil Defense booklets MP-15 (1959), H-12-1 (1980), and H-12-4 (1987). Above: a typical underground concrete shelter (from MP-15).

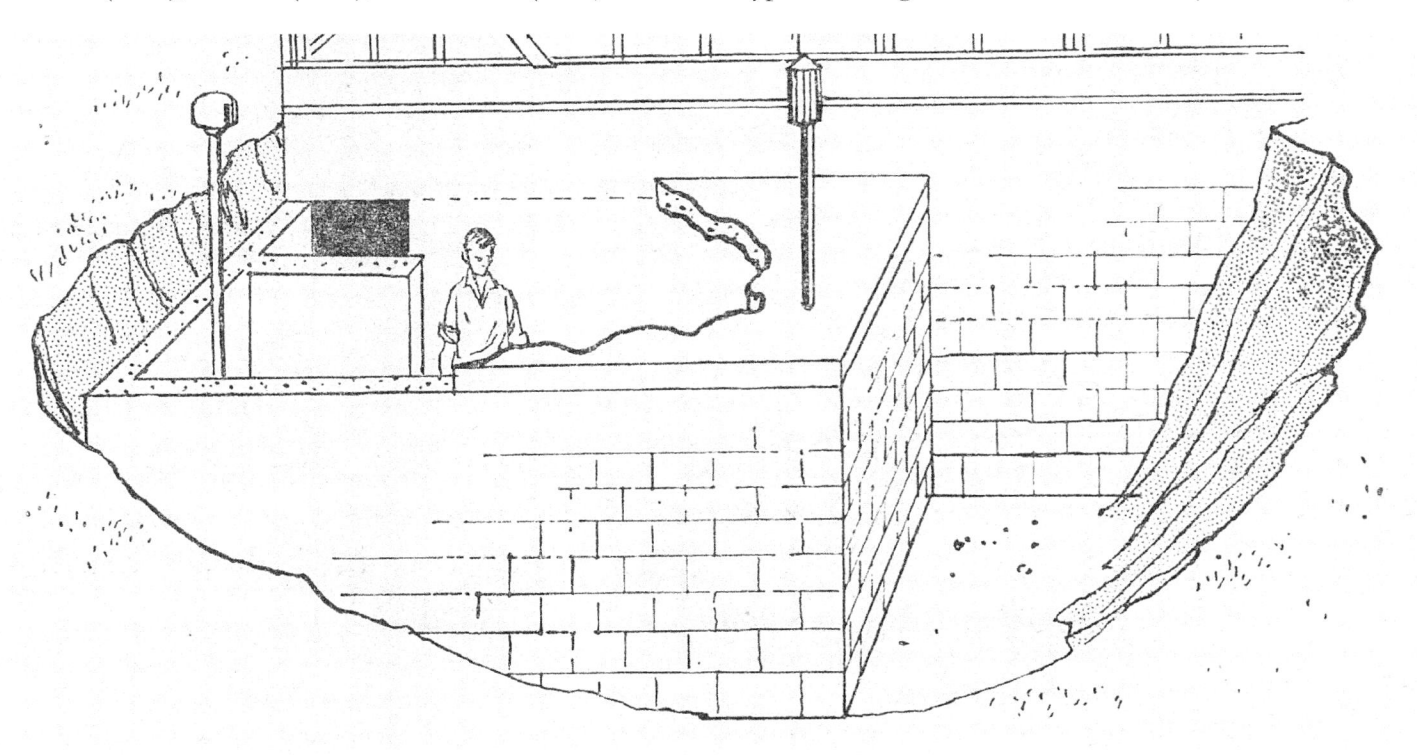

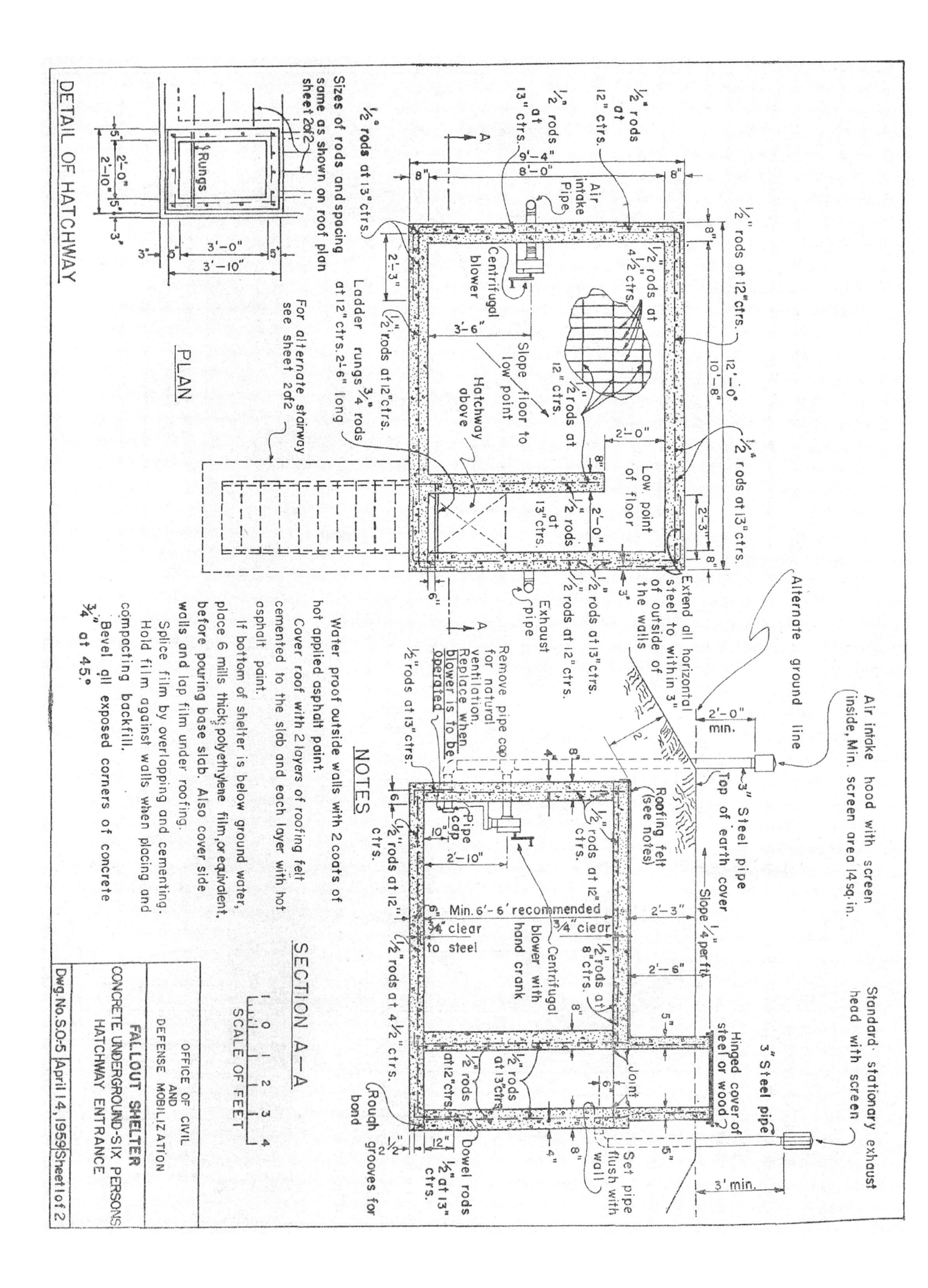

PLAN
(Door Removed)

ROOF PLAN SHOWING REINFORCING

SECTION A-A
ALTERNATE STAIRWAY DETAILS

FOR NOTES:
SEE, SHEET 1 of 2

Same steel sizes and spacings in side walls

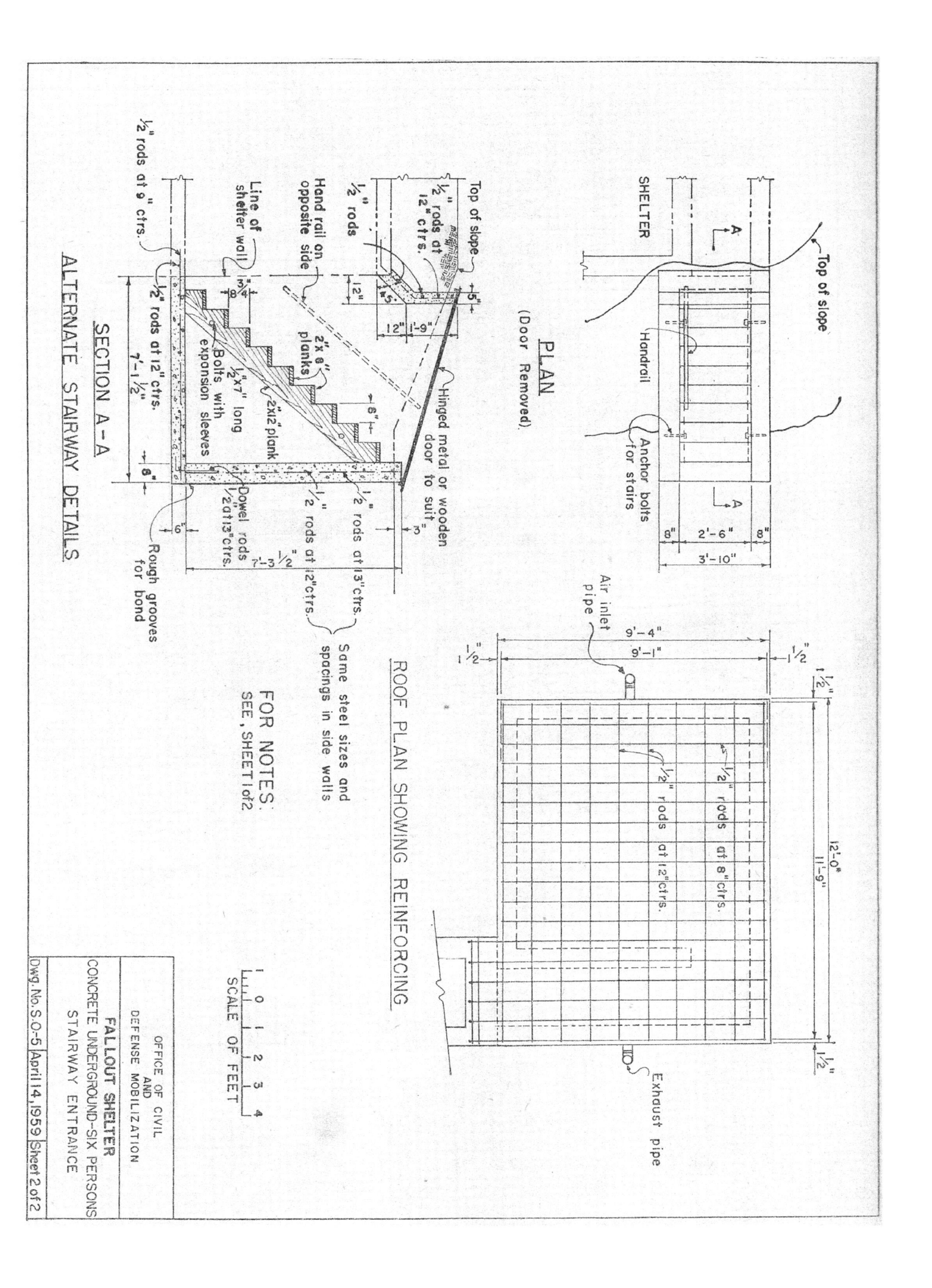

SCALE OF FEET

OFFICE OF CIVIL
AND
DEFENSE MOBILIZATION

FALLOUT SHELTER
CONCRETE UNDERGROUND-SIX PERSONS
STAIRWAY ENTRANCE

Dwg. No. S.O-5 | April 14, 1959 | Sheet 2 of 2

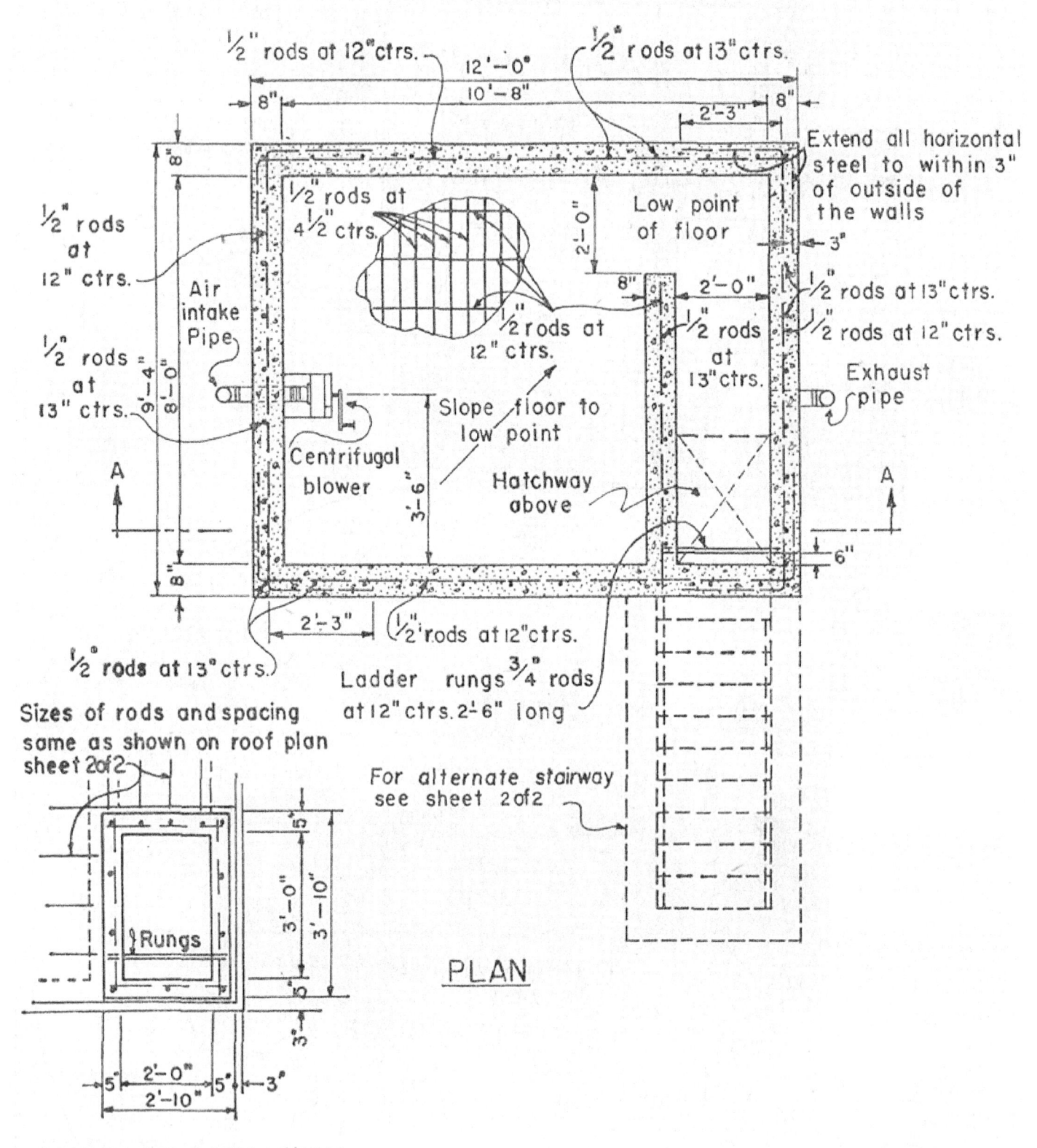

DETAIL OF HATCHWAY

PLAN

Above and right: enlarged views of full plan sheet 1 of 2 (see page 10).

Fallout Shelter Concrete Underground-Six Persons, Hatchway Entrance.
Credit: Office of Civil and Defense Mobilization
Drawing Date April 14, 1959 Publication: MP-15 The Family Fallout Shelter (June 1959)

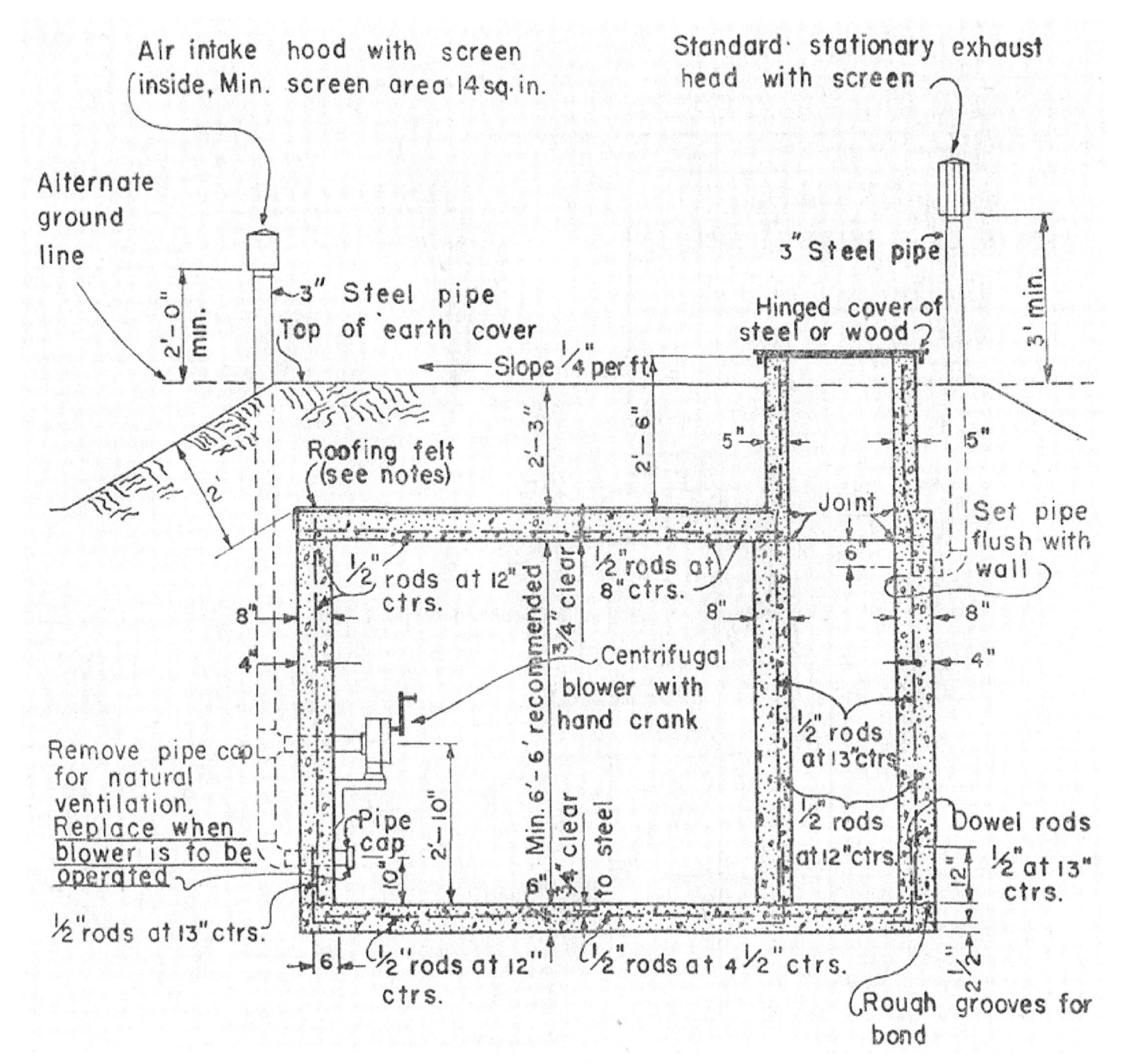

Notes

Water proof outside walls with 2 coats of hot applied asphalt paint*.

Cover roof with 2 layers of roofing felt cemented to the slab and each layer with hot asphalt paint*.

If bottom of shelter is below ground water, place 6 mills thick polyethylene film or equivalent before pouring base slab. Also cover side walls and lap film under roofing.

Splice film by overlapping and cementing. Hold film against walls when placing and compacting backfill.

Bevel all exposed corners of concrete 3/4" at 45 degrees

*Henry 107 Asphalt Emulsion (commonly available in 5 gallon size) can be used in place of hot asphalt paint

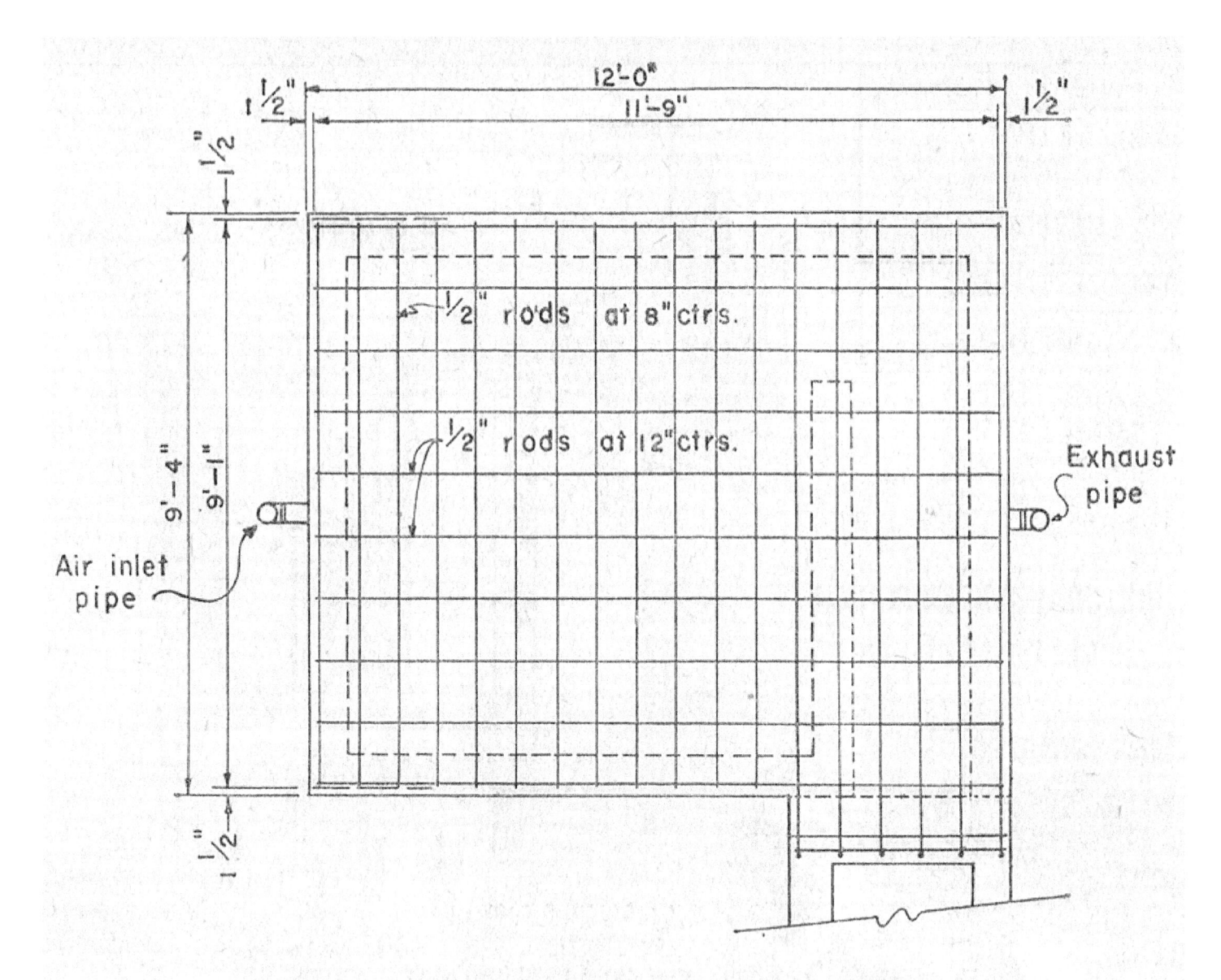

ROOF PLAN SHOWING REINFORCING

Above and right: enlarged views of full plan sheet 2 of 2 (see page 11).

Fallout Shelter Concrete Underground-Six Persons, Hatchway Entrance.
Credit: Office of Civil and Defense Mobilization
Drawing Date April 14, 1959 Publication: MP-15 The Family Fallout Shelter (June 1959)

MP-15 Dimensions

Outside length: 12'-0"   Inside length: 10' 8"

Outside width: 9' 4"   Inside width: 8'-0"

Headroom: 6' 6"   Depth of cover: 27-30"

ALTERNATE STAIRWAY DETAILS

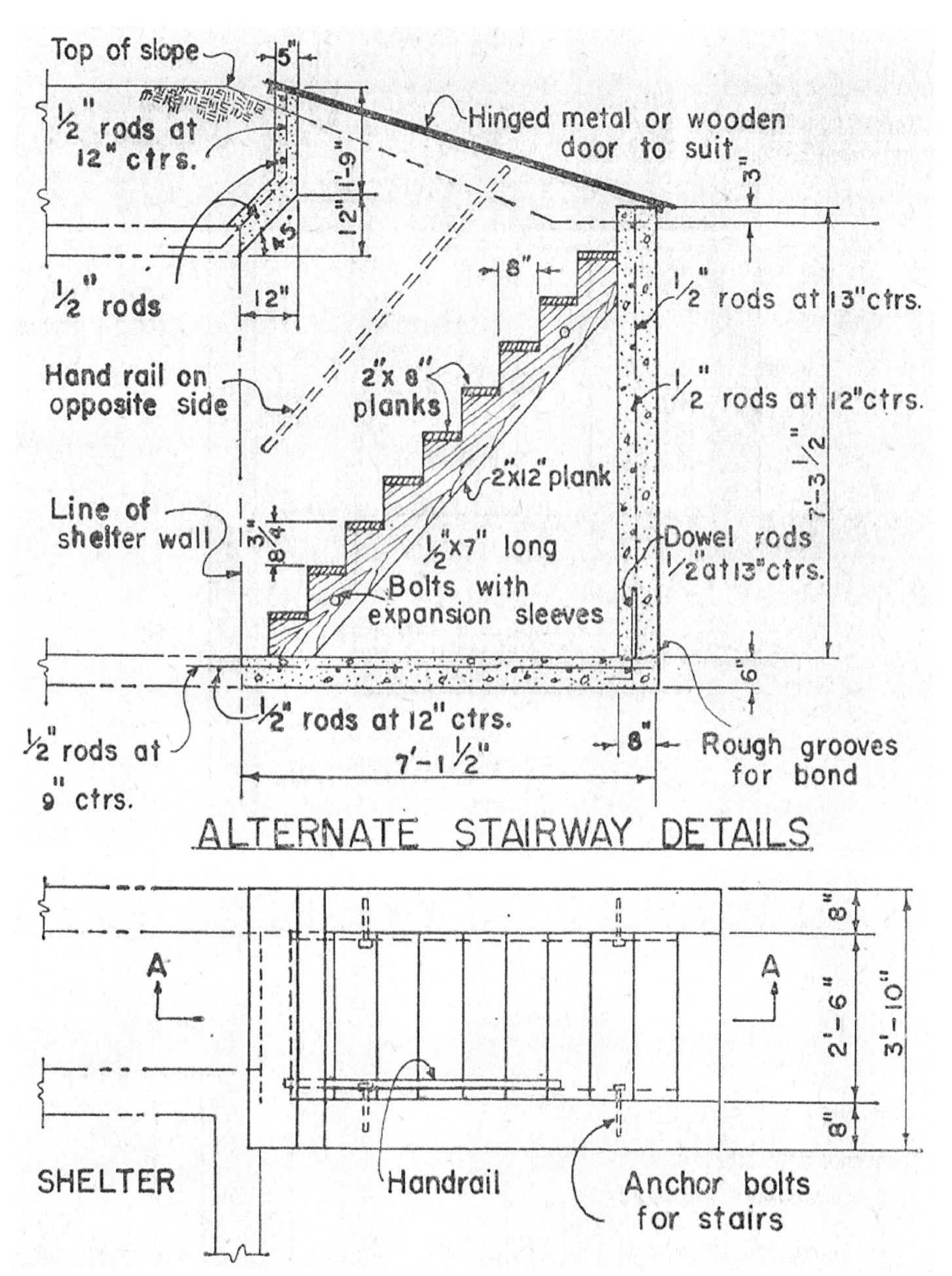

The stand-alone shelter presented in MP-15 is a simple straightforward design with the option for a snorkel type hatchway entrance (shown below) or a more advanced staircase design. As noted before, not everyone is able to navigate a ladder due to physical limitations, or, a staircase might be required for other reasons such as frequent loading and unloading of materials from the shelter. The hatchway and ladder layout offers the benefit of lower cost and easier construction. A tripod and winch can still be used in the hatchway entrance for loading and unloading heavy supplies such as water cans, deep cycle batteries, food, or ammunition. A custom break-down tripod with simple boat-trailer winch is cheap to build, and can be stored inside the shelter when not in use.

If concealing the existence of the shelter is a concern, and I feel this is legitimate, the 3'x4' hatchway entrance presents a slightly smaller (for the purpose of camouflage) upright footprint than the doorway of the larger 6'x3' staircase design (see pages 20-22). Both entrance designs feature the hatchway or door situated 3" from the surface of the ground, perhaps too low to the ground for some builders taste. This design aspect was likely a function to reduce nuclear fallout contact. It can be adjusted by reducing the depth of fill over the shelter as plans call for utilizing a sloped building site, either naturally occurring, or man-made. If building in flat typography, the walls of the hatchway or staircase entrance can be lengthened during the form building process.

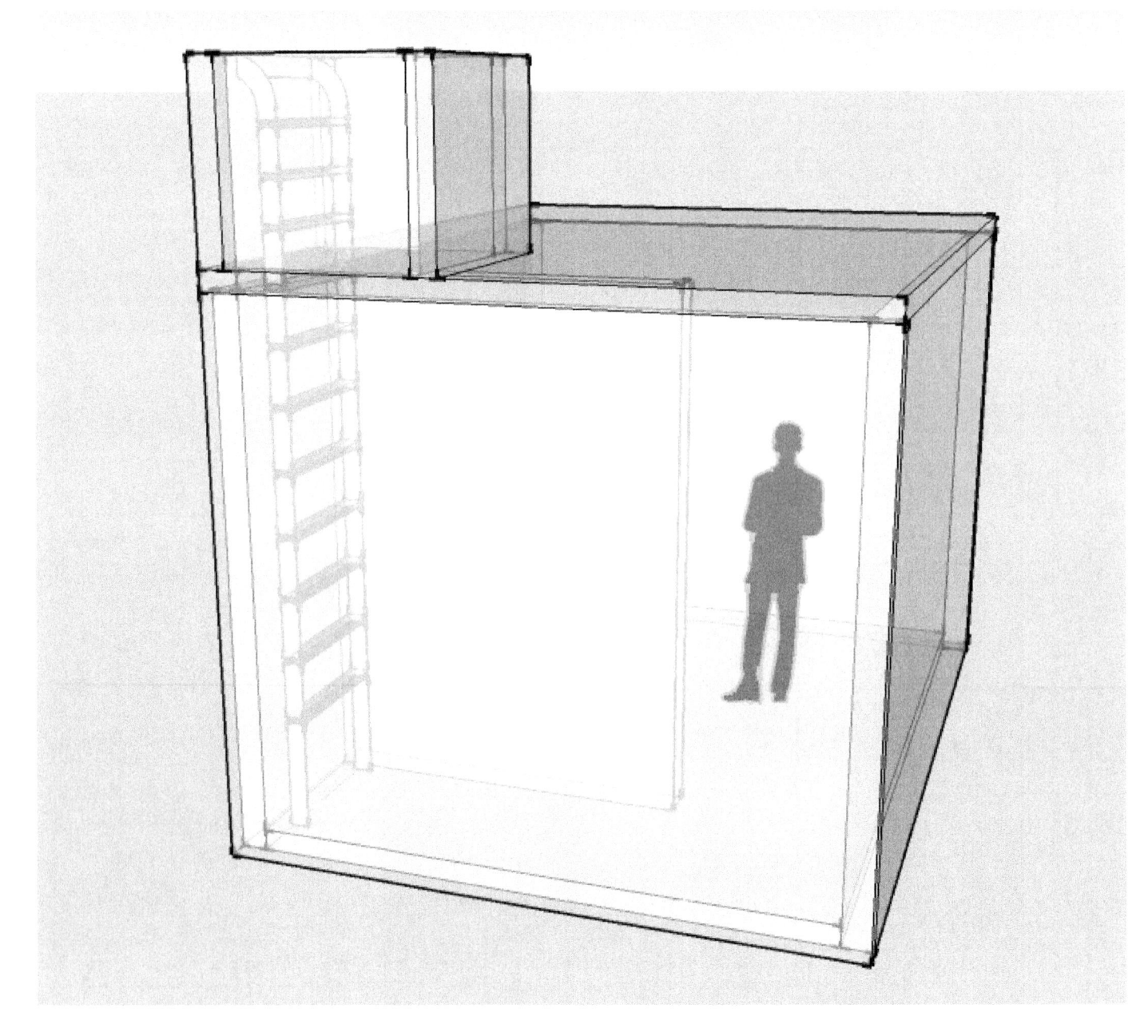

The building of these type shelters is a very similar process to a traditional concrete basement. Extensive excavation and installation of drainage pipes and aggregate are an integral part of the building site preparation process as described on pages 50-53. Ground water levels are also a major concern when considering a true underground shelter. A high water table may cause your excavation to fill with water almost immediately preventing you from using the designs presented in this book. This is not to say it's impossible to pour a concrete shelter in a high water table area, but it might not be worth all the inevitable moisture problems you will encounter after it's completed. See pages 54-61 for Aboveground Masonry Block Bomb Shelter Plans.

Even in areas with low water tables, water will naturally be drawn toward a hole in the ground. Storms bringing heavy rain can lead to water infiltration due to power failures that render effective sump pit and pump designs useless. Most of these problems can be avoided with proper design and planning. Back-up power sources like a deep cycle battery bank kept charged by a solar panel, for example, could power a sump pump for several days depending on the number of batteries in the bank.

When in doubt always use your local builders basement construction methods as a general guideline of successfully built underground rooms, the vast majority of which never experience major water infiltration problems.

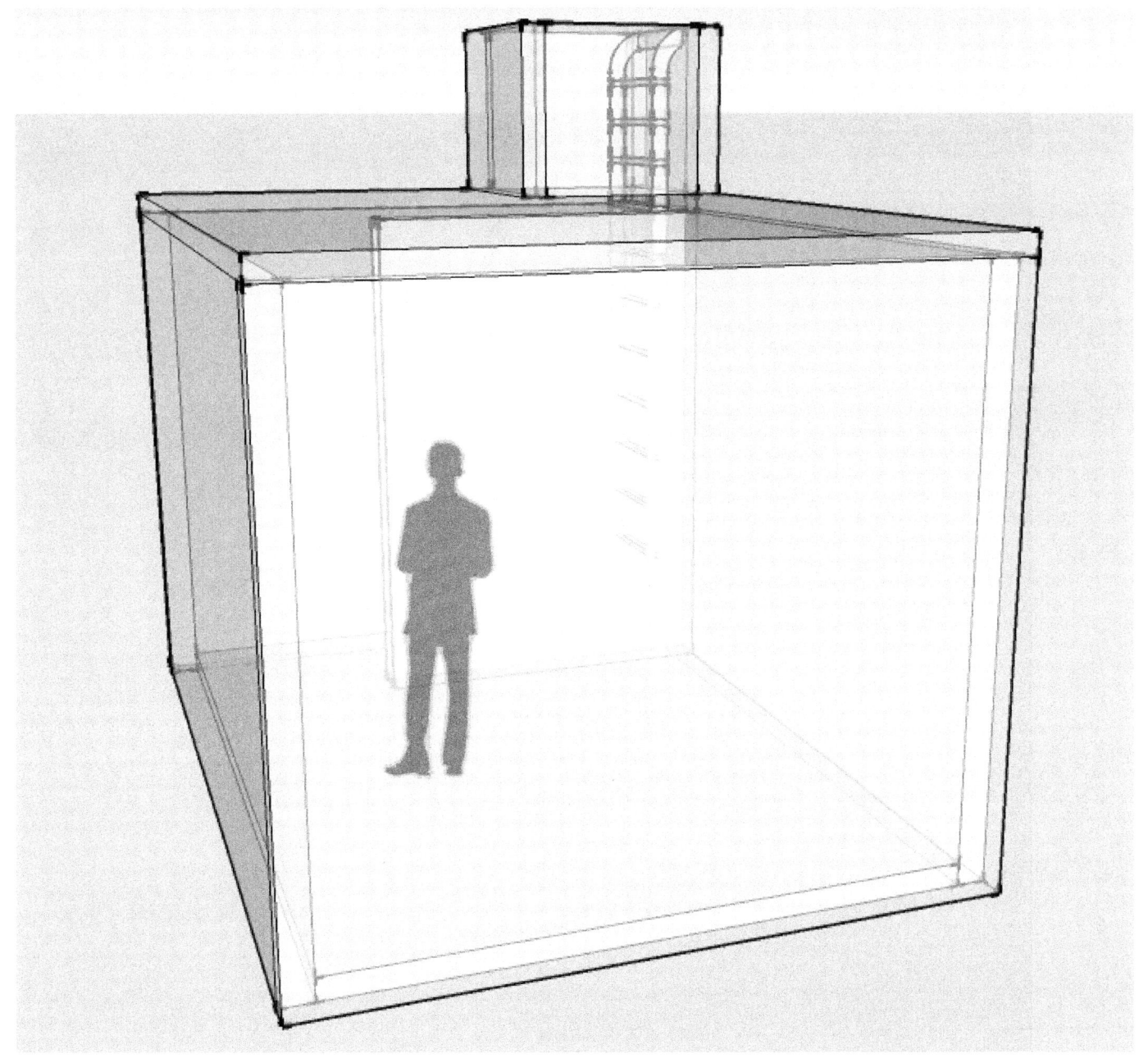

Ladder Fabrication - Fallout Shelter Concrete Underground-Six Persons, Hatchway Entrance, MP-15.

The hatchway designed underground shelter utilizes a vertical ladder for gaining entrance and exits. The illustration labeled 'detail of hatchway' on page 12 shows steels rungs embedded directly in the concrete wall. This method will produce a streamlined ladder using a minimal amount of material, but preparations must be made in the concrete form prior to the pour. Placing temporary 1" pvc segments wrapped in bubble wrap in the form will create holes for anchoring the ladder rungs. Use a bender to shape pieces of #5 (5/8") rebar into ladder rungs and paint each with a heavy duty epoxy paint. Use anchoring cement to permanently embed the rebar rungs into the pre-formed mounting holes that remain after the pvc segments are knocked-out.

Alternately, you can fabricate a steel ladder similar to the one above with mounting brackets for anchor bolts at the top, two near mid-span, and bottom. Inside width of hatchway passage is listed at an even 2 ft, with the length ranging from 7-9 ft depending on mounting height.

While you are having your ladder fabricated, and also during the actual building of the shelter... a standard hardware store type aluminum extension ladder can serve as a fully functional temporary ladder.

Fallout Protection Factor - The protection factor for a shelter of this type is over 1000 meaning it will reduce gamma ray exposure by a factor of 1000.

Ventilation - Ventilation equipment and pipe are required. A hand operated blower should be specified to furnish at least 20 cubic feet of air per minute. The air is exhausted through a separate vent pipe (see page 13).

Interior - It is suggested that the interior walls and ceiling be finished with two coats of light colored cement paint, and that the floor in the shelter area be finished with light colored tile or latex base paint.

Furnishing - Furnishings should be of the type that maximize space such as fold-up bunks, or modular loft bunks with dining or desk area below. Space saving boat cabin and RV designs can be mimicked in the shelter by installing beds, stoves, galleys, and overhead storage bins from boat and RV new supply, used, or salvaged sources.

Water - Store water in 5 gallon water cans (commonly these are blue plastic jugs). Keep a gallon of plain household bleach on hand for disinfecting drinking water if required. Use 8-15 drops of bleach per gallon, 1 teaspoon of bleach per 5 gallon can. Safe bleach-disinfected water will be relatively clear with a slight odor of chlorine.

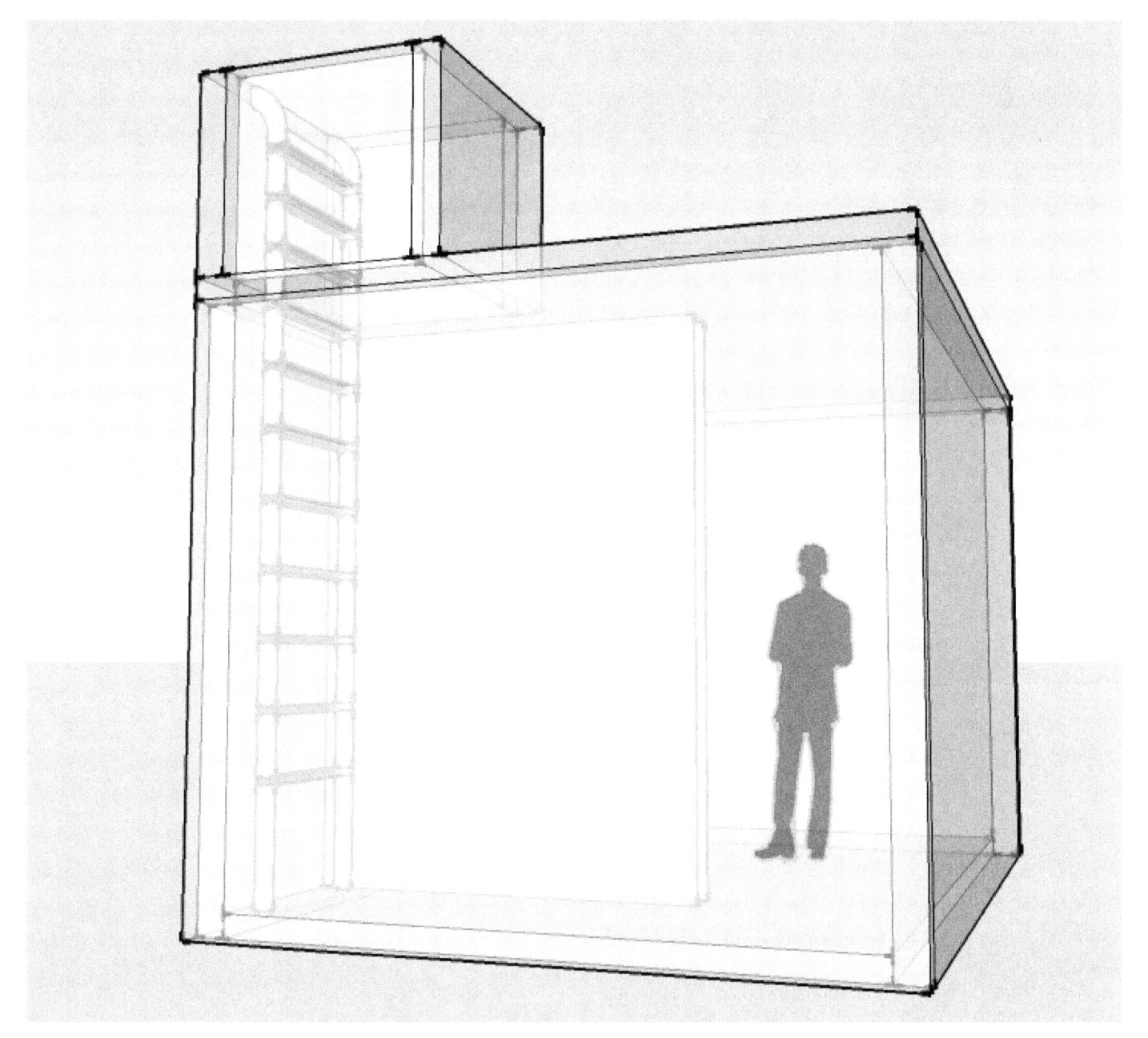

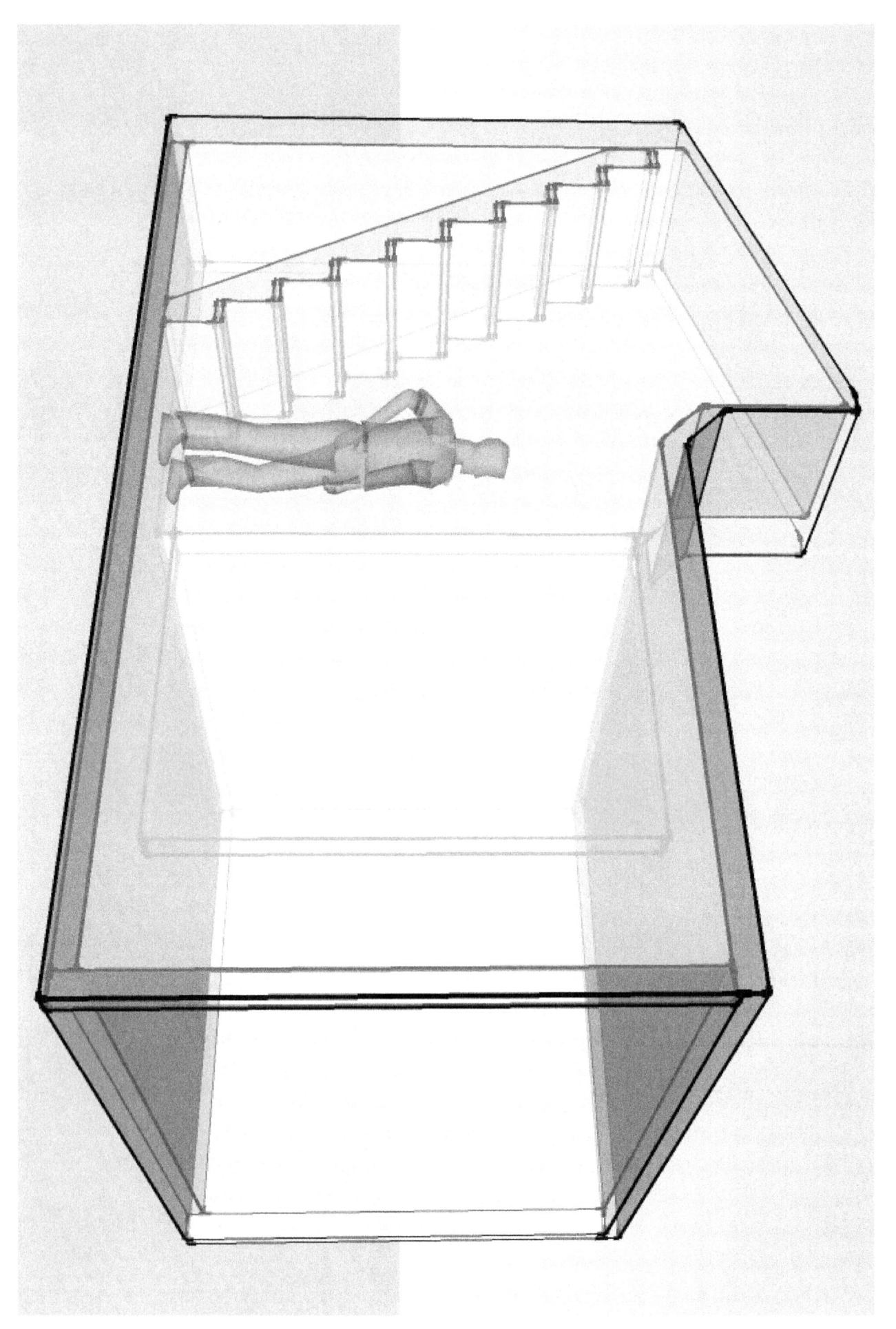

Alternate stairway variation of MP-15 underground shelter.

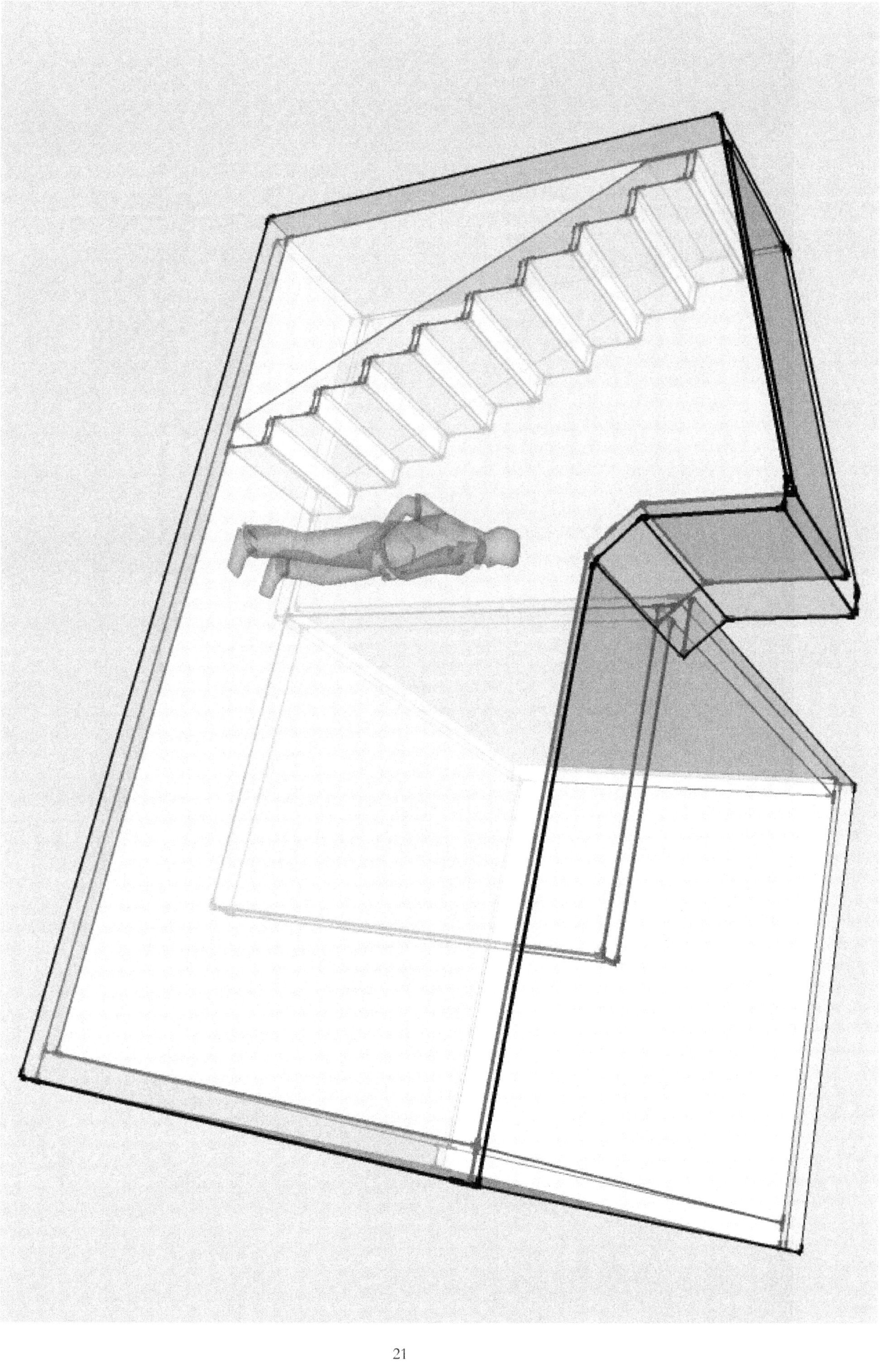

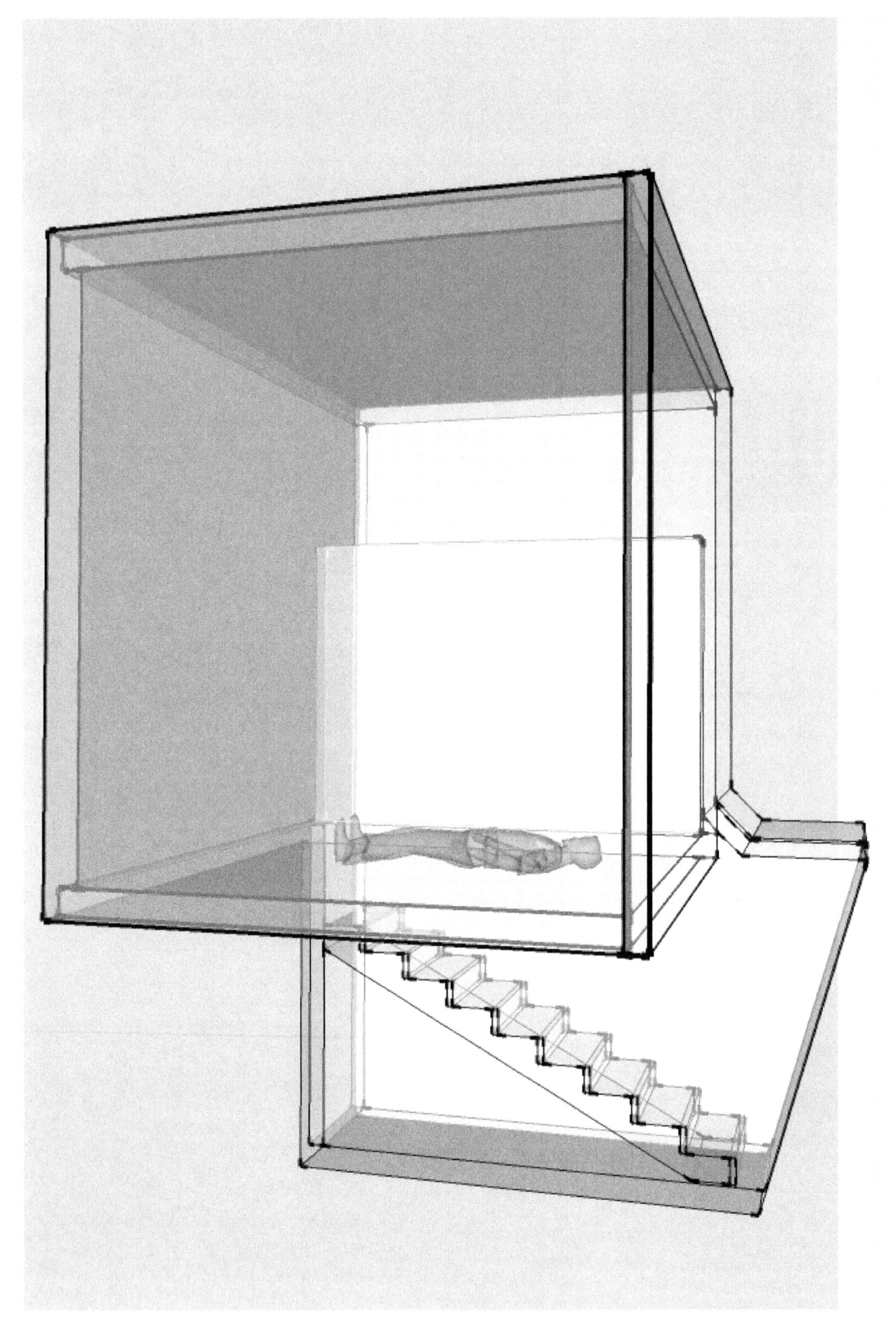

Alternate stairway variation of MP-15 underground shelter (details on page 15).

Aspects of H-12-1 Home Shelter Plans Booklet (1980)

The shelter presented in this booklet distributed in 1980 is essentially a slightly updated version of the plans contained in the MP-15 booklet (seen on previous pages). There are a few changes in dimension.

The MP-15 shelter plan has an outside dimension of 12' x 9'-4" (6'-6" headroom).

The H-12-1 shelter plan has an outside dimension of 12'-8" x 9'-4" (6'-8" headroom).

There is an option to accommodate additional persons by increasing the shelter length by 2'-6" for each two people. The booklet notes not to increase the 9'-4" width as this will change the roofs load-bearing dynamics.

Another difference with this newer design is it's shallower excavation. The roof placement is level and in-line with grade allowing the actual top of the slab roof to be used as a patio.

The booklet also gives several options for modifying the basic plan where the depth of cover is 24" like the original MP-15 design. My feeling is the shallow shelter excavation is better suited as a tornado shelter, nevertheless, the booklet states this shelter is designed to provide a fallout protection factor of 40.

Excerpts from H-12-1 (continued on pages 24-25) ...

Protection is provided in an outside concrete shelter. The roof of the shelter can be used as an attractive patio.

General Information

This family fallout shelter, designed primarily for homes without basements, is a permanent home shelter to be placed in the yard. It is designed to have a protection factor of at least 40, which is the minimum standard of protection for public shelters throughout the United States. This assures that persons inside the shelter will be protected against radioactive fallout following a nuclear attack, and will also have some protection against blast and fire effect of nuclear explosions.

The shelter is capable of housing six adults. It can be built of poured reinforced concrete, precast concrete slabs, or a combination of concrete blocks and poured concrete. If it is built as detailed with the top near ground level, the roof slab can be used as an outdoor patio. The shelter is accessible by a hatch-door and wood stairway. Fresh air is provided by a hand-operated centrifugal blower and two ventilating pipes that extend above ground level. In areas where there is poor drainage or where the ground water table is close to the surface, the fourth modification on page 5 should be used. Before starting to build the shelter, make certain that the plan conforms to the local building code. Obtain a building permit if required. If the shelter is to be built by a local contractor, engage a reliable firm that will do the work properly and offer protection from any liability or other claims arising from its construction.

Excavation

The excavation should have side slopes gradual enough to prevent caving, or appropriate shoring should be provided. Materials used for backfill and embankment should have debris, roots and large stones removed before placement. The subgrade for the floor slab should be level for ease in placing waterproofing membrane and to provide uniform bearing conditions for the structure. The area surrounding the patio should be sloped away at a minimum grade of 1 inch per 10 feet to provide good drainage.

This shelter will withstand overpressures of up to 5 psi, and provides excellent protection from tornadoes.

H-12-1 Home Shelter

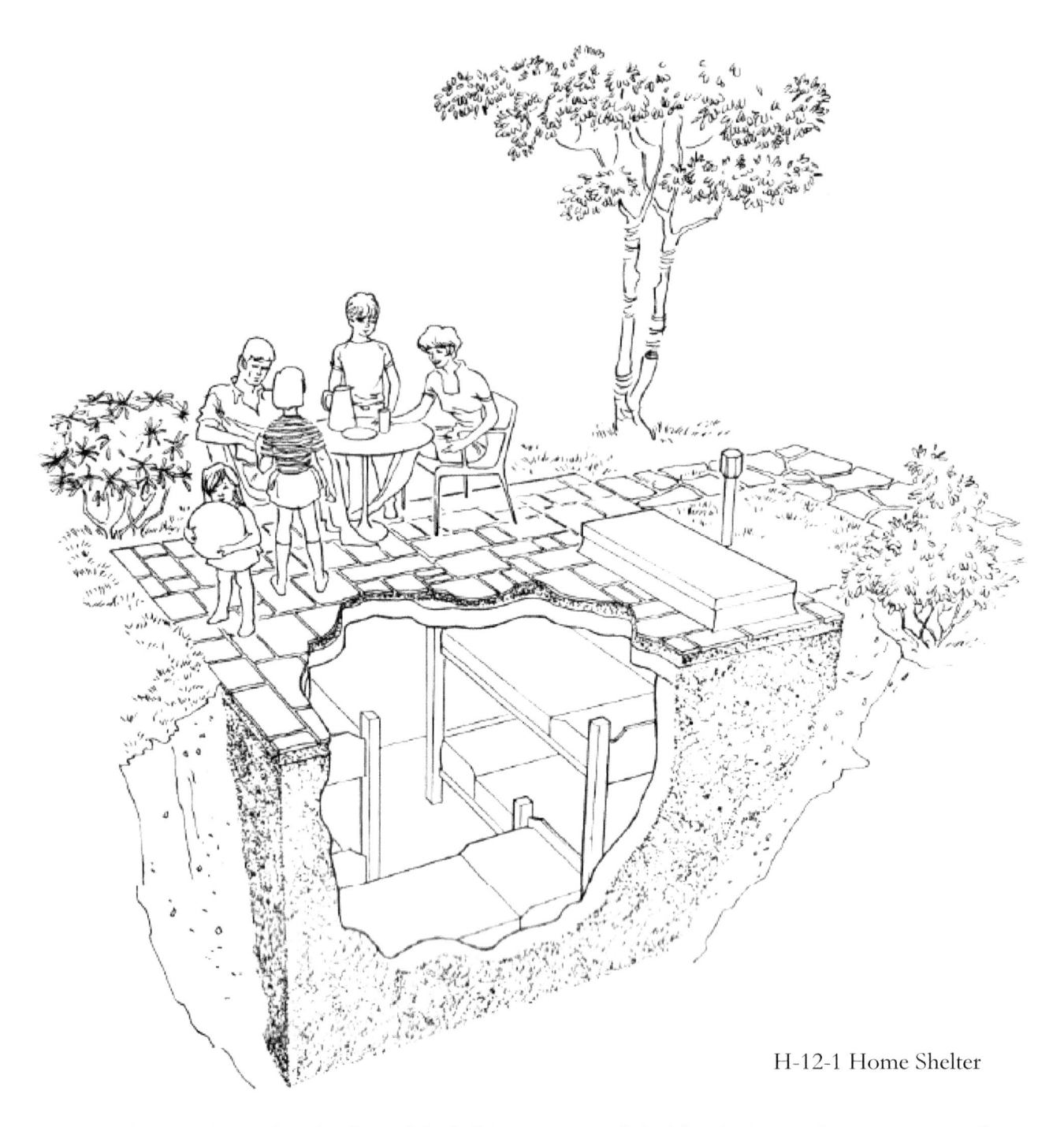

Exterior walls, roof slab and under floor slab shall be waterproofed with a 3-ply membrane waterproofing system. This provides a continuous blanket which seals the entire area of surface to be protected. The membrane shall be protected from backfill damage and when completing other stages of construction. Place flagstone or bricks on a sand bed when using the roof slab as a patio. There are a number of commercially produced metal roof hatches that will adequately serve as a shelter door. However, as long as the door is weatherproof and durable, a job-made, galvanized sheet metal covered wood door is suitable. Bevel all exposed corners of concrete 3/4" at 45'.

<u>Structural Design Data</u>

Steel = 20,000 psi

Concrete = 2,500 psi

Soil (minimum) = 600 psf, to withstand downward pressure

Modifications

This first modification utilizes 12-inch concrete masonry units for walls instead of reinforced concrete. The floor, roof and entranceway are the same as in the basic shelter, and the amount of protection provided is essentially the same.

If a basement is available, the shelter may either be separate from it, or attached. In this modification, an attached shelter is entered through the basement of the house, thereby permitting dual use of the shelter space. Other advantages of this modification include flexibility of shape and design to conform to the house design and the use of the same kind of building materials as used in the construction of the house.

If the topography permits, the shelter can be built into a hillside or embankment. This modification increases the protection factor by the addition of an earth mound over the shelter. A maximum of 3 feet of earth cover is recommended.

Material List

Structure	Concrete Requirement	Structure	Steel Reinforcing
Floor	60 cu ft	Floor	580 lineal feet
Walls	235 cu ft	Walls	945 lineal feet
Roof	50 cu ft	Roof	260 lineal feet
Total	345 cu ft / 13 cu yds	Total	1785 lineal feet

Miscellaneous Materials

(2) tie wire - 6" coils

(1) hand blower w/mounting bracket

(16 lin. ft) 3" galv. steel pipe

(2) 3" galv. ells

(1) 3" galv. tee

(1) 3" galv. cap

(1) intake hood, w/screen

(1) exhaust hood, w/screen

(2) wood carriages, 2" x 12" x 10'

(9) wood treads, 2" x 8" x 2'-8"

(2) wood plates, 2" x 4" x 2'-8"

(1) hatch door, metal covered

(1) wood plate, 2" x 8" x 7'

(1) wood plate, 3" x 8" x 14'

(3) T-hinges, 8" x 5-1/2" E. H., galv.

(1) hasp and staple, galv.

(1) chain door stop, galv.

(8) anchor bolts, 1/2"+ x 8"

(4) expansion shields and bolts, 3/8" Q x 4"

(715 sq. ft) waterproofing membrane

(100 sq. ft) flagstone

(1.5 cu. yds) sand

(12 lin. ft) cant strip

H-12-4 (1987) Home Shelter: Plans for a home shelter of masonry block that provides protection from nuclear fallout radiation and tornadoes.

Plans for a concrete-block built design were also made available by The Civil Defense Administration and FEMA. The most all-encompassing design was from booklet H-12-4 (1987) Home Shelter: Plans for a home shelter of masonry block that provides protection from nuclear fallout radiation and tornadoes.

The main benefit of using CMU's (concrete masonry units) is the elimination of the need for forms which equals a major reduction in cost and labor. Concrete block in conjunction with rebar, mortar, and corefill is a combination that has been used extensively and successfully to build many thousands of residential basements during the past century seen frequently in post WW2 home building.

The drawbacks of concrete block is a higher incidence of cracks and water infiltration over the life of the shelter. Laying concrete blocks also requires some level of skill. You can only use concrete block to build the walls, meaning... concrete slabs for the foundation and roof will still need to be poured. These will require building support forms. Nevertheless, there will still be a significant savings on concrete delivery and form lumber.

Note that there are different grades of concrete block, some types are manufactured from lighter density materials at a lower cost. This plan calls for standard dimension premium quality all-concrete CMU block.

See page 48 for more information on building with concrete masonry units.

The concrete block built shelter shown in booklet H-12-4 is essentially the same size as the concrete shelters previously described compared below (outside dimensions) :

MP-15 shelter plan is 12' x 9'-4" (6'-6" headroom).
H-12-1 shelter plan is 12'-8" x 9'-4" (6'-8" headroom).
H-12-4 shelter plan is 12' x 9'-4" (6'-8" headroom).

Excerpts from H-12-4 ...

Description of the Shelter

This protective shelter is designed to serve as a family fallout shelter and is suitable for other utilitarian purposes as well including use as a tornado shelter and everyday functions of the residence. The shelter is designed for placement in the yard and primarily is for houses without basements. To function as a fallout shelter, it is designed to have a protection factor (PF) of at least 40, which is the minimum standard of protection for family and public shelters recommended by the Federal Emergency Management Agency (FEMA). The belowground location of the shelter also will provide some protection against blast and fire effects of a nuclear explosion.

The structure may also be used as a highly effective tornado shelter in regions of the nation where they occur. The roof structure of the shelter is suitably strong to resist the most severe tornadoes, and the belowground location provides protection for occupants from wind blown debris and even from possible collapse of nearby buildings.

The day-to-day use for the particular design illustrated is for housing residential swimming pool filtration equipment in a weather protected location out of sight in the yard. Other utilitarian uses for the facility are possible that may be more suited to a particular homeowner's needs such as use for yard equipment storage or use as a cellar for storage of perishable foods or the facility may be used solely as a refuge from natural and man-made hazards.

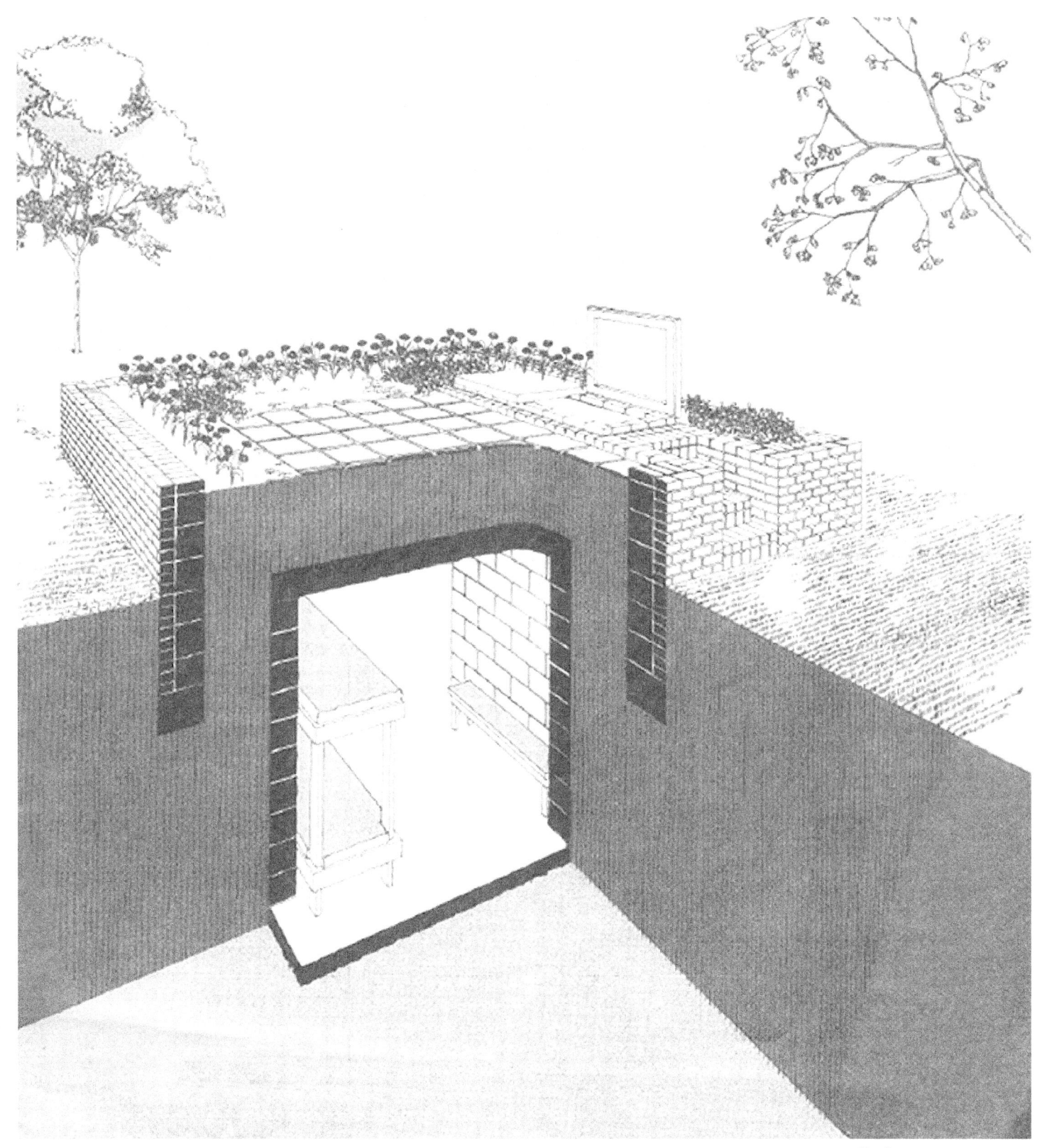

H-12-4 Home Shelter

H-12-4 is unique in that it utilizes a shallow excavation where the top of the slab roof is in line with the natural ground level. There is a 2ft high, 17ft x 14ft retaining wall surrounding the perimeter of the shelter allowing for 2 ft depth of back-fill cover. The original plan specifies this wall to be constructed of concrete block with a decorative layer of clay brick over it creating an attractive 1980's style patio and landscaped structure which contains steps to the hatchway entrance of the shelter. Specialized retaining wall concrete blocks with a stackable design (as described on page 49) have since become commonly available at home improvement stores and could be substituted for both the standard CMU's and clay bricks resulting in a considerable savings in labor and expense.

Excerpts from H-12-4 (continued)

An elevated brick planter placed atop the roof of the shelter creates a landscape feature in the yard and provides overhead protection against radioactive fallout and tornado forces. Attractive landscaping and enhanced protection are achieved with this arrangement without burying the shelter deeply into the ground. Other sitting arrangements besides the raised planter are possible for the basic shelter. For example, the brick planter walls can be eliminated and the concrete roof slab can be paved as a terrace at the yard level. Elimination of the soil cover atop the shelter will reduce, but not make ineffective, the overall fallout radiation protection of the space. Whatever landscape treatment may be preferred, the plans shown in this booklet are valid for construction of the basic shelter. The facility also can serve as a storm shelter.

Plans for the Shelter

Plans illustrated in this booklet are for a shelter to accommodate up to six adults. The shelter has reinforced concrete floor and roof slabs and reinforced masonry block walls. An elevated planting area, with brick faced garden walls, retains a 2ft. deep soil cover over the roof of the shelter. Access to the shelter is by means of a hatchway and wood stair. Provisions are made for ventilating the shelter space by means of a hand-operated centrifugal blower. Air intake and exhaust pipes extend above the ground level of the planter.

Dimensioned plans in this booklet provide sufficient information for a professional contractor to build the shelter. For the novice "do-it-yourself" builder, a companion booklet, H-12-4.1, is available from the Federal Emergency Management Agency that provides step-by-step instructions plus additional details for construction of the shelter. A list of construction materials is provided on the back page of this booklet. It includes quantities for all materials needed to complete the construction except miscellaneous items such as stakes, nails, and other fasteners. The companion booklet, H-12-4.1, provides more detailed information on sizes and quantities of materials needed for each phase of the construction.

Note: I contacted the FEMA library regarding this booklets availability… they told me that the H12 series of booklets described here are no longer available. I also searched extensively for used copies online over the period of a year without managing to locate a single original printed copy of the H-12-4 series of booklets. If you have one hang onto it, they are rare.

Building the Shelter:

Layout and Excavation

Initial layout of the shelter entails the measuring and marking necessary to correctly locate the facility in the yard. Care should be exercised in this phase of the work to assure that the shelter will be built where it is intended and at the depth intended in relationship with yard elevations. Side walls of the excavation should be sloped sufficiently so that soil will not slough off into the work area. Alternatively, the side walls can be shored if the soil is especially loose.

During the excavation phase, do not excavate deeper than the bottom level of the slab or drainage fill (if any) to assure that the bearing soil is not disturbed.

Footings and Floor Slab

A combined footing and floor slab is designed for the shelter. By thickening the slab at its edges, support is provided for the block walls. Underground utilities should be placed before the floor slab is poured - such as a floor drain or sump pit, and water piping. A sump is provided for floor drainage of the shelter illustrated, but other drainage methods can be used provided that the drainage water has someplace to flow.

All concrete should have a minimum compressive strength of 2,500 lbs. per sq. inch (psi). Locations and sizes of reinforcement steel are indicated in the plans. All reinforcement indicated in the plans should be installed even though it might seem possible to omit some.

Masonry Block Walls

The walls of the shelter are constructed of standard 8" thick masonry block (CMU's). The block walls are reinforced both horizontally and vertically. Prefabricated trussed wire reinforcement is used in horizontal joints, placed continuously at every second bed course. Vertical reinforcement is No.4 steel bars spaced at 8" on centers (one bar in each block cell). Every other vertical bar (one each block unit) is secured to dowels formed in the floor slab.

Cells of the block units are grouted to provide a bond between vertical reinforcement steel and block units. Grout lifts should not be greater than 4 feet for any one pour. Type S mortar is specified for block masonry joints and for grout.

Roof Slab

A reinforced concrete roof slab 8" thick is designed for the shelter. The roof slab is supported on the masonry block walls. Sizes and locations of reinforcement steel for the roof slab are described on page 39.

Reinforcement consists of No.4 bars spaced 8" O.C. running in the direction of the short dimension of the shelter (structural reinforcement) and No.4 bars spaced at 16" in the long dimension (temperature reinforcement). Reinforcement around the hatchway opening, also No.4 bars, is indicated in the plans. Accommodations for sump discharge pipe, or any electrical conduits not routed through walls, must be made in the slab form.

Damp-proofing / Waterproofing

Protection of the underground facility from water and moisture penetration is recommended. If soil conditions are relatively dry and if there is good surface water drainage, then damp-proofing should be sufficient. If ground water is observed in the excavation and if the excavation is likely to become a collector basin for water, then waterproofing probably will be necessary to achieve a dry shelter space. Damp-proofing and waterproofing concepts and techniques are described on pages 13, 50-53.

Planter Walls

Walls of the surrounding planter are constructed after the basic shelter is completed and after backfill is placed up to a level of the footings for the planter walls. Concrete footings for the planter walls should be set below the frost line depth for the region where the shelter is built.

Excerpts from H-12-4 (continued)

Planter walls consist of 4" standard face brick and 8" backup block. The planter walls are capped with brick. Type S mortar also is specified for this work.

These walls do not require grouting or vertical reinforcement bars unless the height of the walls above the surface of the yard is greater than about 3 feet. Horizontal joint reinforcement for a 12" wall should be used in alternate bed courses of the block.

Brick steps leading to the hatchway are indicated in the plans for the shelter.

Ventilation

The ventilation system for the shelter consists of a hand-operated centrifugal blower, air intake pipe with filter hood, and air discharge pipe with hood. Air intake and discharge pipes are placed on opposite or adjacent walls of the shelter space to provide optimum movement of ventilation air. Piping should be placed with outlets more or less at the heights above the floor level of the shelter as shown in the plans. Piping and fittings may be either galvanized steel or ABS (plastic). The air intake pipe is fitted with a hood and screen filter so that radioactive particles will not be pulled into the shelter space by the blower. The air exhaust pipe is hooded, but no filter is needed.

Centrifugal blowers can be purchased commercially.

Note: Most modern models are powered by an electric motor requiring electric service, or a battery bank and inverter. You could also modify a modern electric centrifugal blower to be dual-use (electric and manual) by removing the motor and installing a custom made removable hand crank, making it useable if power is lost (see original description below).

Electrical service for lighting and power equipment also may be added. Electrical service should be from a separate circuit and with a branch circuit breaker inside the shelter that has ground fault protection. An electrically powered centrifugal blower may be substituted for the hand-operated blower. It should be recognized that electrical power to the shelter may be disrupted by a tornado or nuclear explosion. Air intake filter and exhaust hood can be fabricated by a local sheet-metal shop in the homeowner's area in accordance with details included in the plans.

Modification of Plans

The shelter plans shown on subsequent pages may be modified within certain limitations as may be necessary to meet particular needs of the homeowner. To accommodate more than six occupants, increase the length of the shelter 2'-8" for each two additional occupants. The width of the shelter should not be increased unless the roof structure is redesigned. The roof structure of the shelter illustrated is designed to span in the short dimension, and new engineering analysis is needed for longer spans.

Other designs for an elevated planter and for access into the shelter are possible without changing the basic shelter. Each homeowner's preference for landscape character can be met in this phase of the work. Piping for water may be added during construction such as for the swimming pool filter equipment that is illustrated.

List of Construction Materials

Material	Item	Grade	Size	Quantity
Formwork Lumber**	Roof Slab, Footings	Construction	2" x 10"	184 lineal ft.
	Shoring	Construction	2" x 6"	70 lineal ft.
	Edge Forms, Shoring	Construction	2" x 4"	24 lineal ft.
	Batter Boards	Construction	1" x 4"	24 lineal ft.
	Plywood	C-D	1/2" x 4' x 8'	4 pieces
Concrete	Footing / Floor Slab	2500 psi	—	2.75 cu. yds.
	Roof Slab	2500 psi	—	2.75 cu. yds.
	Planter Footings	2500 psi	—	2.00 cu. yds.
Reinforcement	Deformed Bars	Grade 40	No. 4	1134 lineal ft.
	Wire Fabric	ASTM A185	6 x 6 - 8 / 8	108 sq. ft.
	Joint Reinforcement	Trussed Wire	For 8" Wall	228 lineal ft.
	Joint Reinforcement	Trussed Wire	For 12" Wall	130 lineal ft.
	Prefab. Corner Reinf.	Trussed Wire	For 8" Wall	20 pieces
	Prefab. Corner Reinf.	Trussed Wire	For 12" Wall	8 pieces
Masonry	Block	Concrete	8" Standard	638 units
	Block	Concrete	4" Standard	10 units
	Brick	—	Standard	2100 units
	Mortar	Type S	—	57 cu. ft.
	Grout	Type S	—	94 cu. ft.
Hatchway Cover	Lumber	Construction	2" x 4"	18 lineal ft.
	Lumber	Construction	3" x 4"	6 lineal ft.
	Plywood	C-D	1/2" x 4' x 8'	1 piece
	Edge Trim	No. 2 Pine	1" x 4"	24 lineal ft.
	Sheet Metal	Galvanized	26 Ga.	28 sq. ft
	Hinges	—	—	4 pieces
Stair	Strings	No. 2 Fir	2" x 12"	24 lineal ft.
	Treads	No. 2 Fir	2" x 8"	20 lineal ft.
Ventilation	Centrifugal Blower	—	—	1 unit
	Wall Sleeves	Steel or ABS	8" Long	3 pieces
	Piping	Steel or ABS	3" ID	22 lineal ft.
	90 deg. Elbows	Steel or ABS	3" ID	2 pieces
	Tees	Steel or ABS	3" ID	1 piece
	Intake Filter	—	—	1 unit
	Exhaust Hood	—	—	1 unit
Miscellaneous	Screws, Misc. Fasteners,			—
	Plumbing (Optional)			—
	Wiring (Optional)			—

** Some forming lumber may be reused during different phases of the construction work.

Plan showing planter walls and hatchway walls.

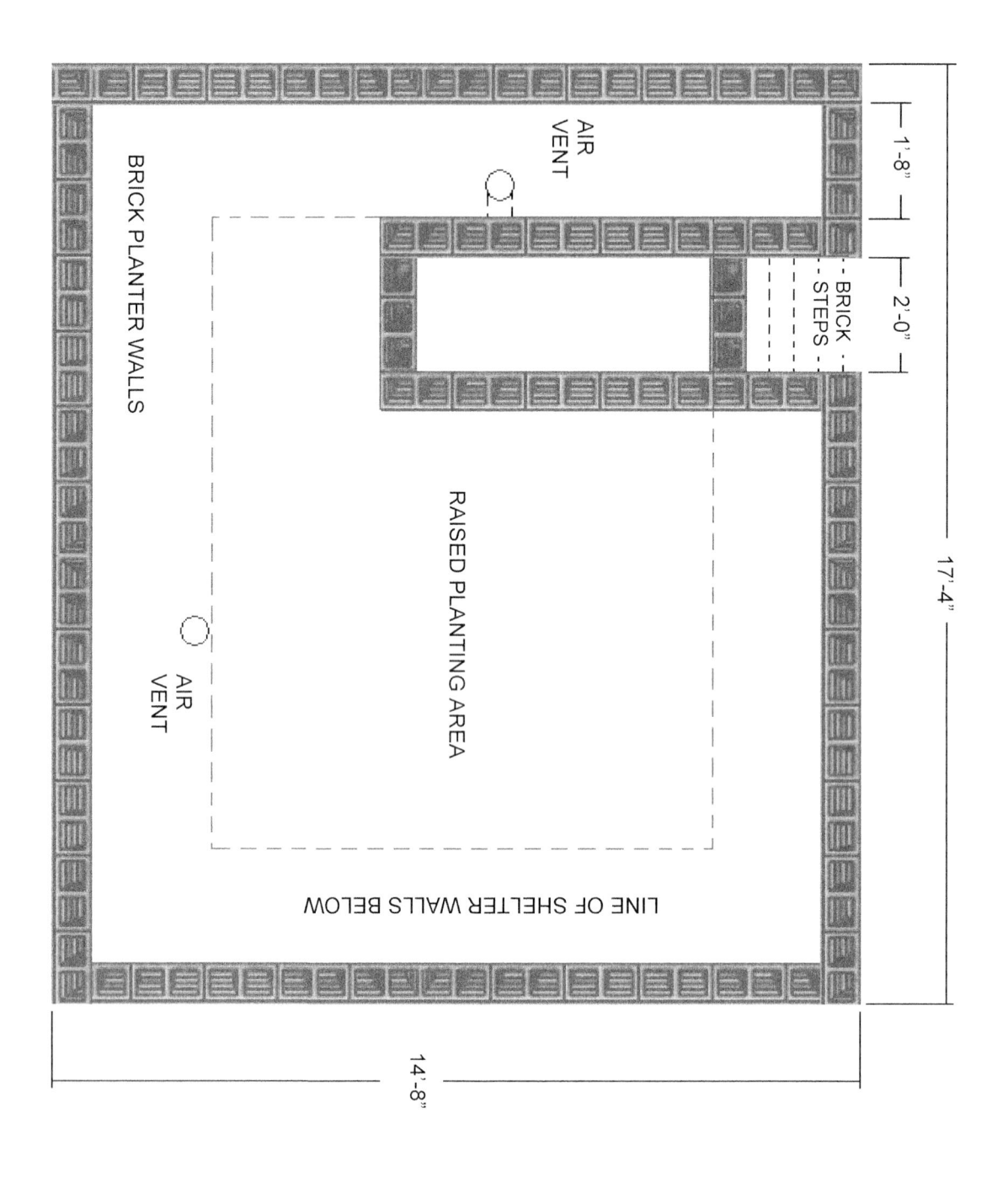

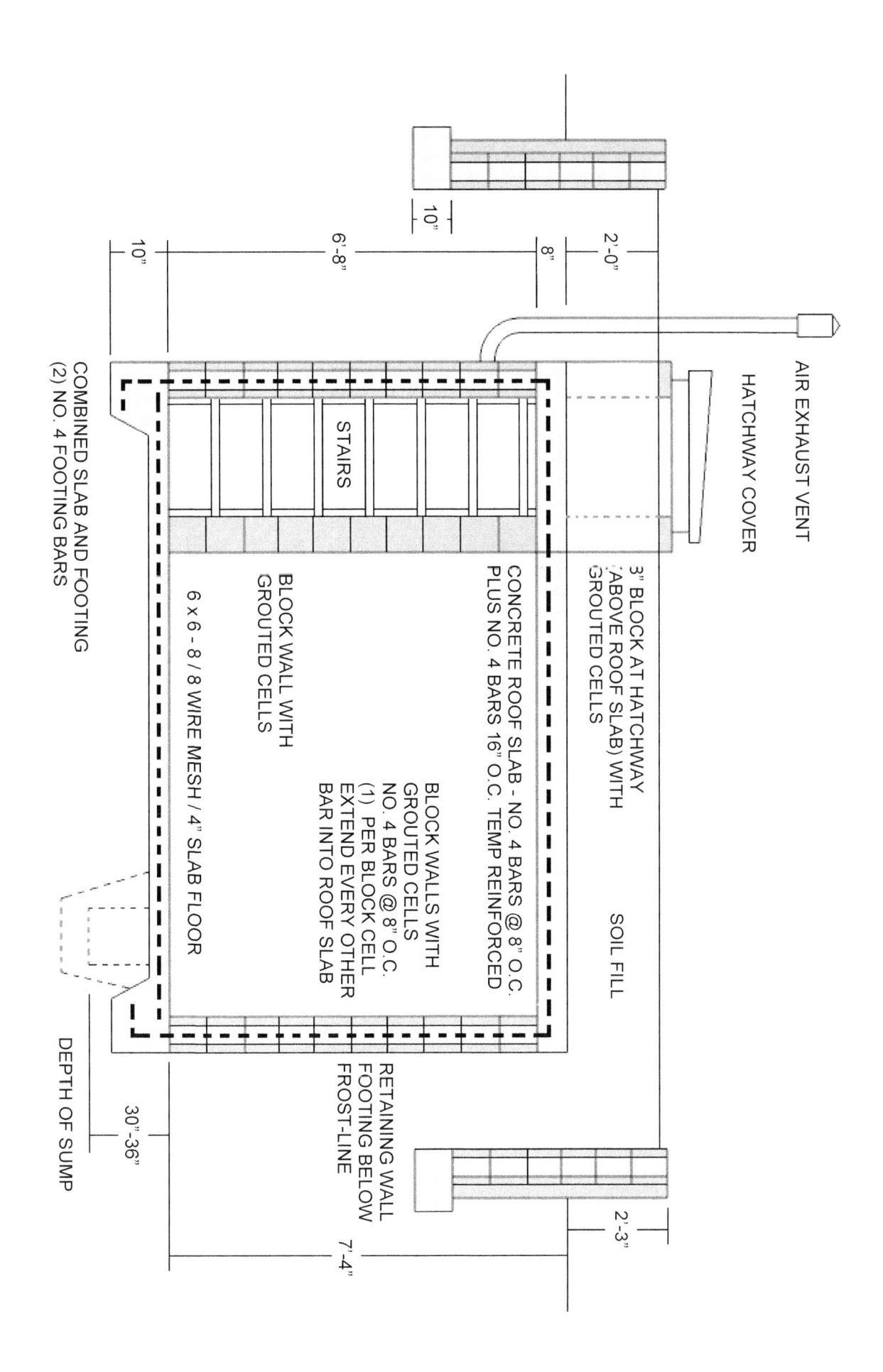

Cross section A view of shelter.

Floor plan of the shelter.

34

Cross section B view of shelter.

Air intake hood detail.

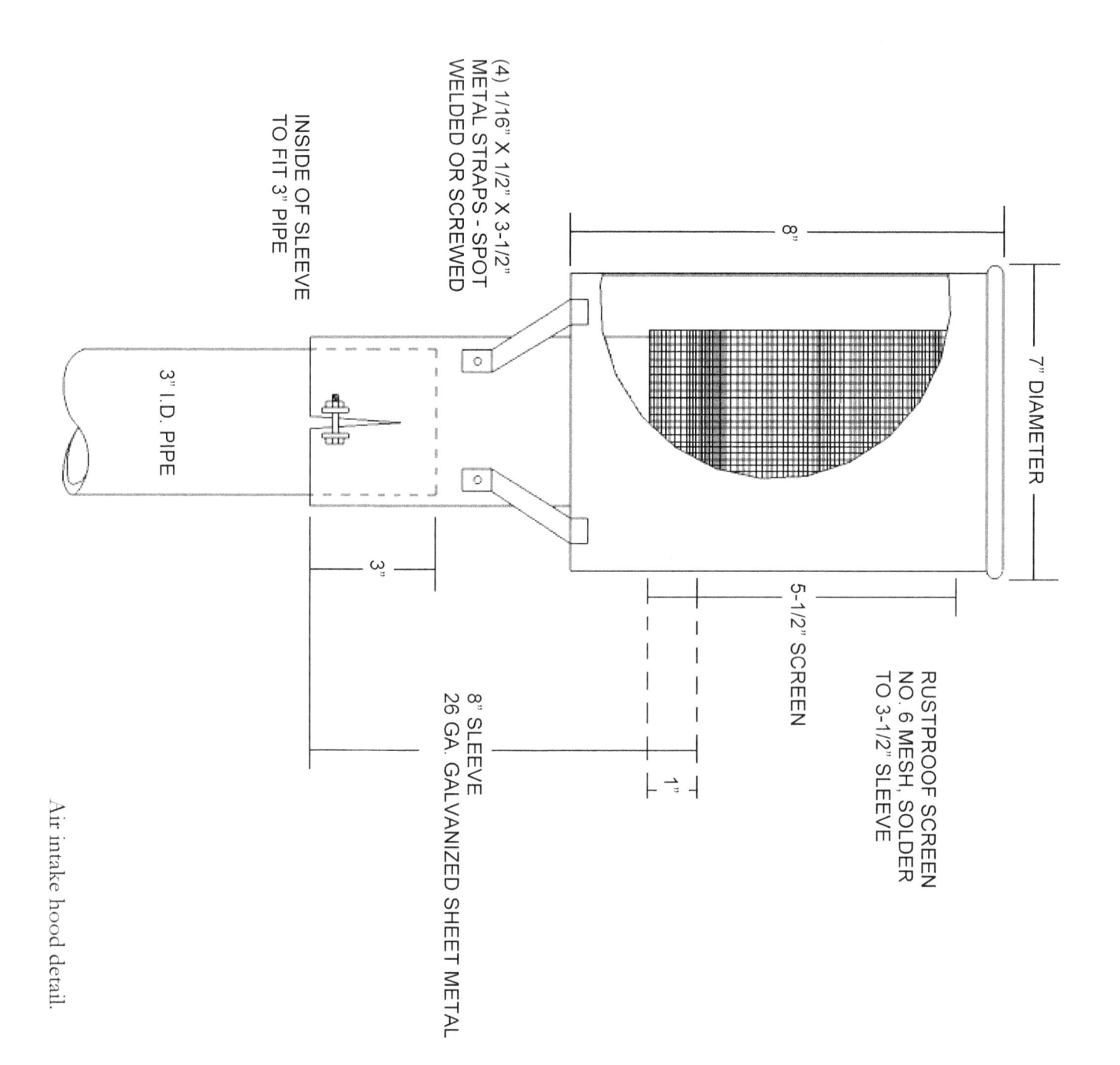

Cross section C view of shelter at hatchway.

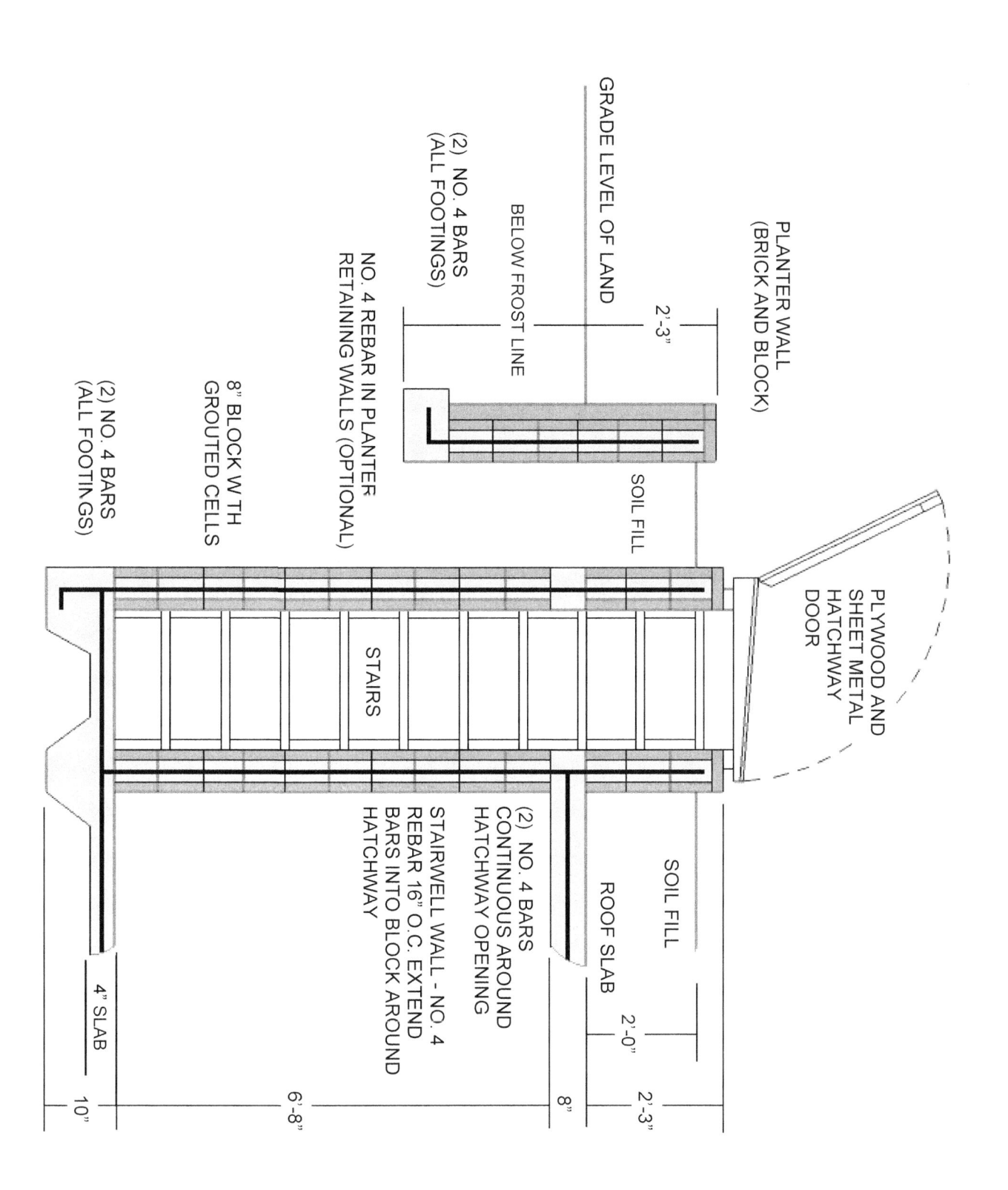

Section D view at hatchway.

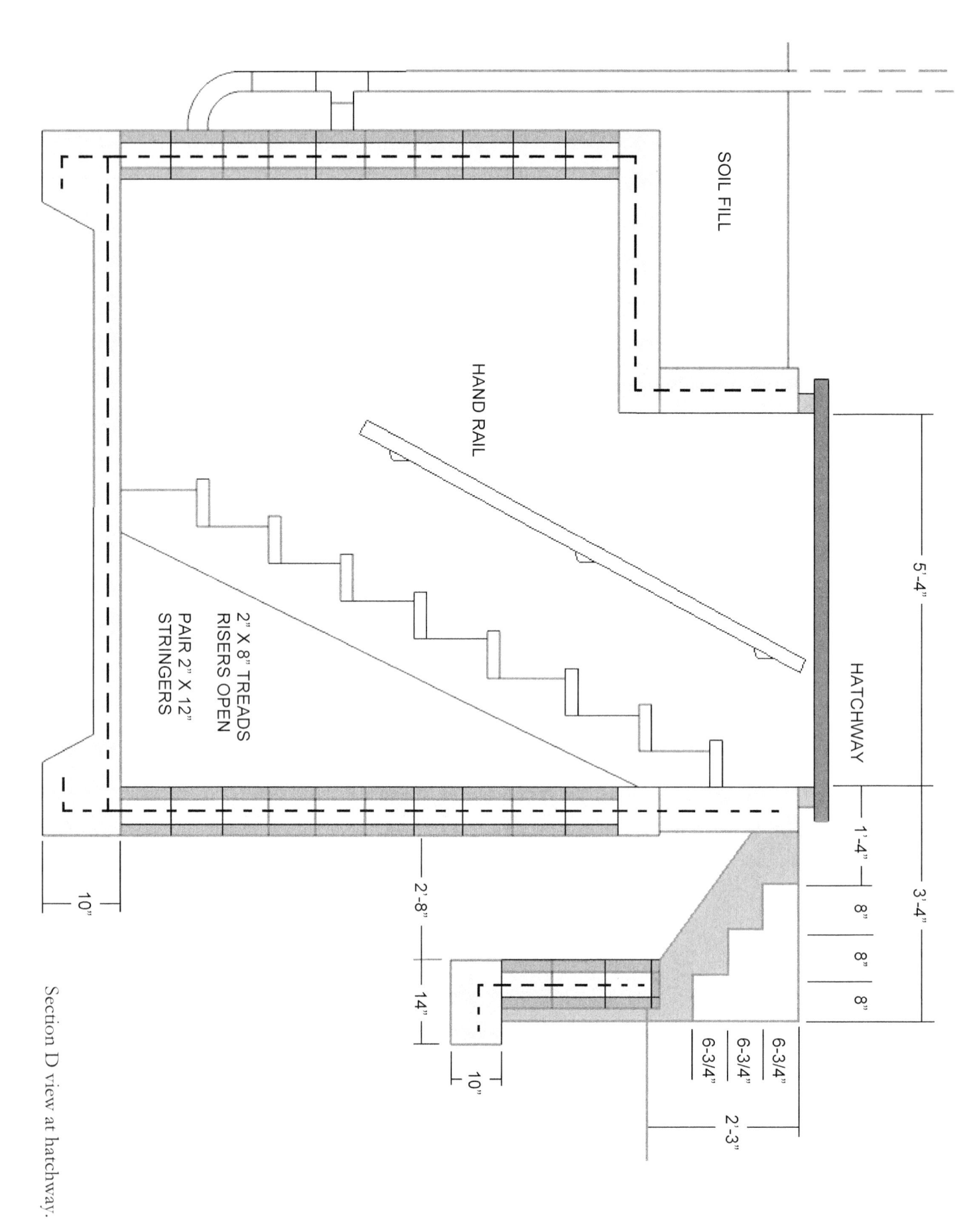

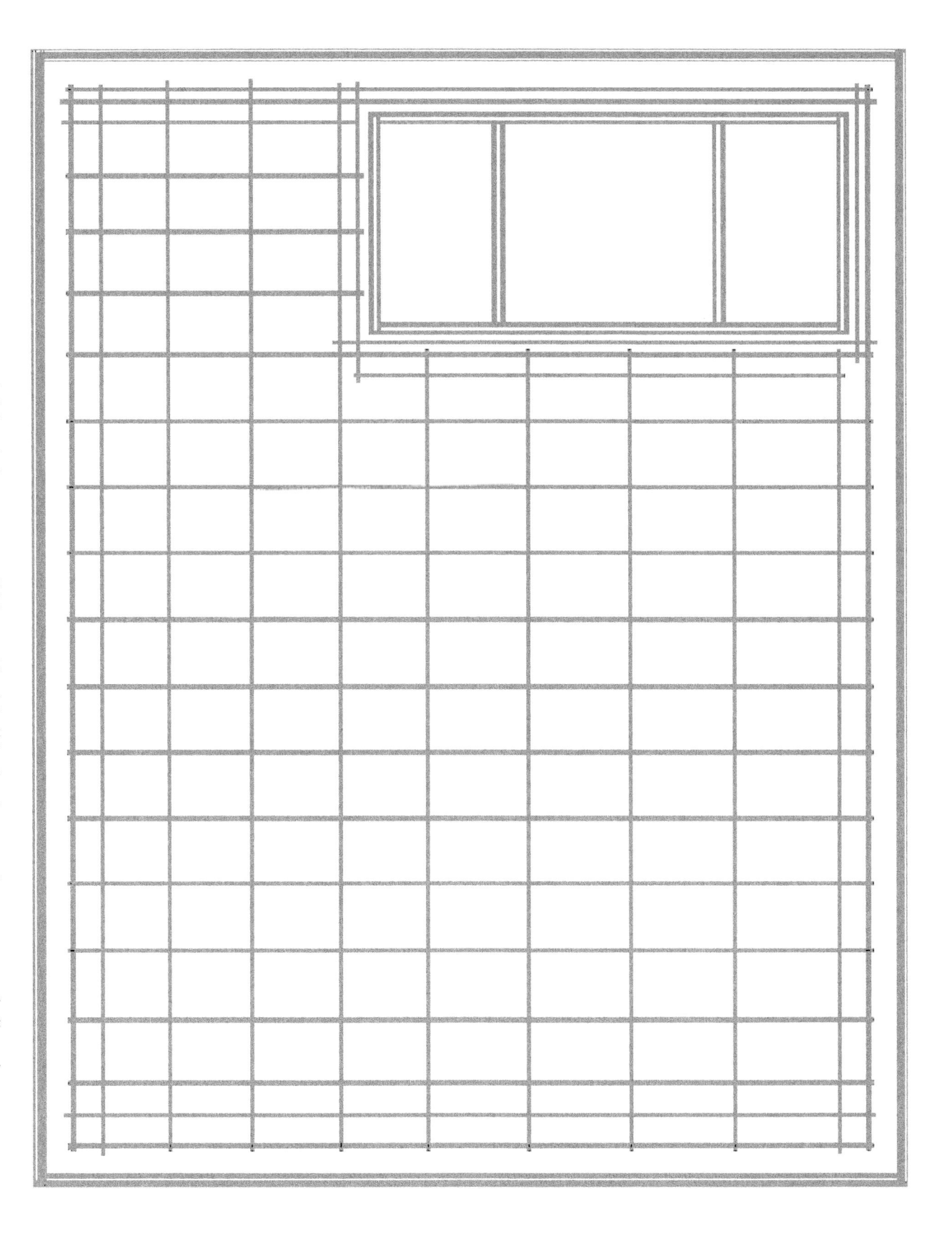

MP-15 and H-12-4 concrete roof slab reinforcement: #4 bar grid with #4 edge bar continuous around all perimeters.

2'-10-1/2"

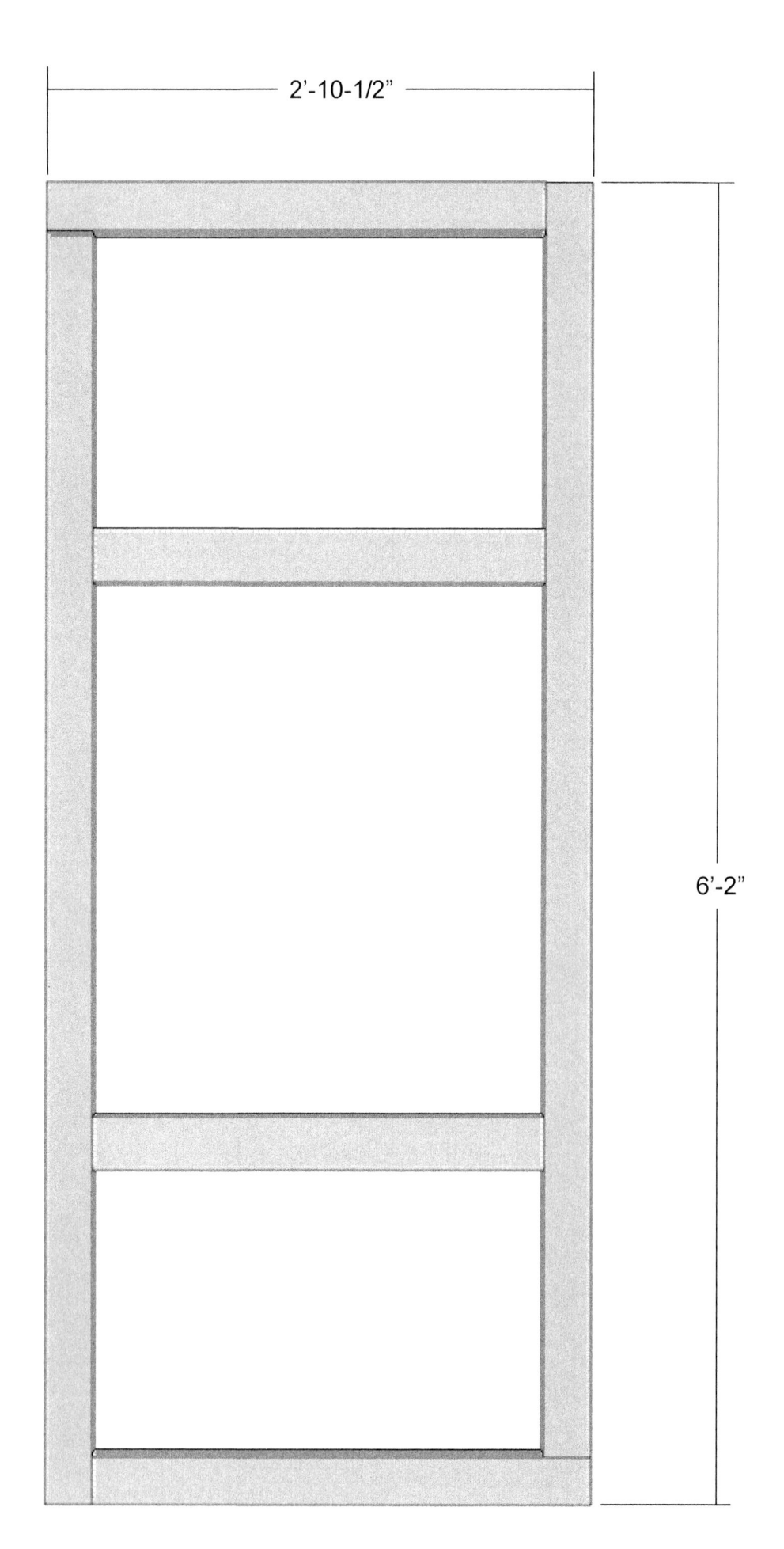

6'-2"

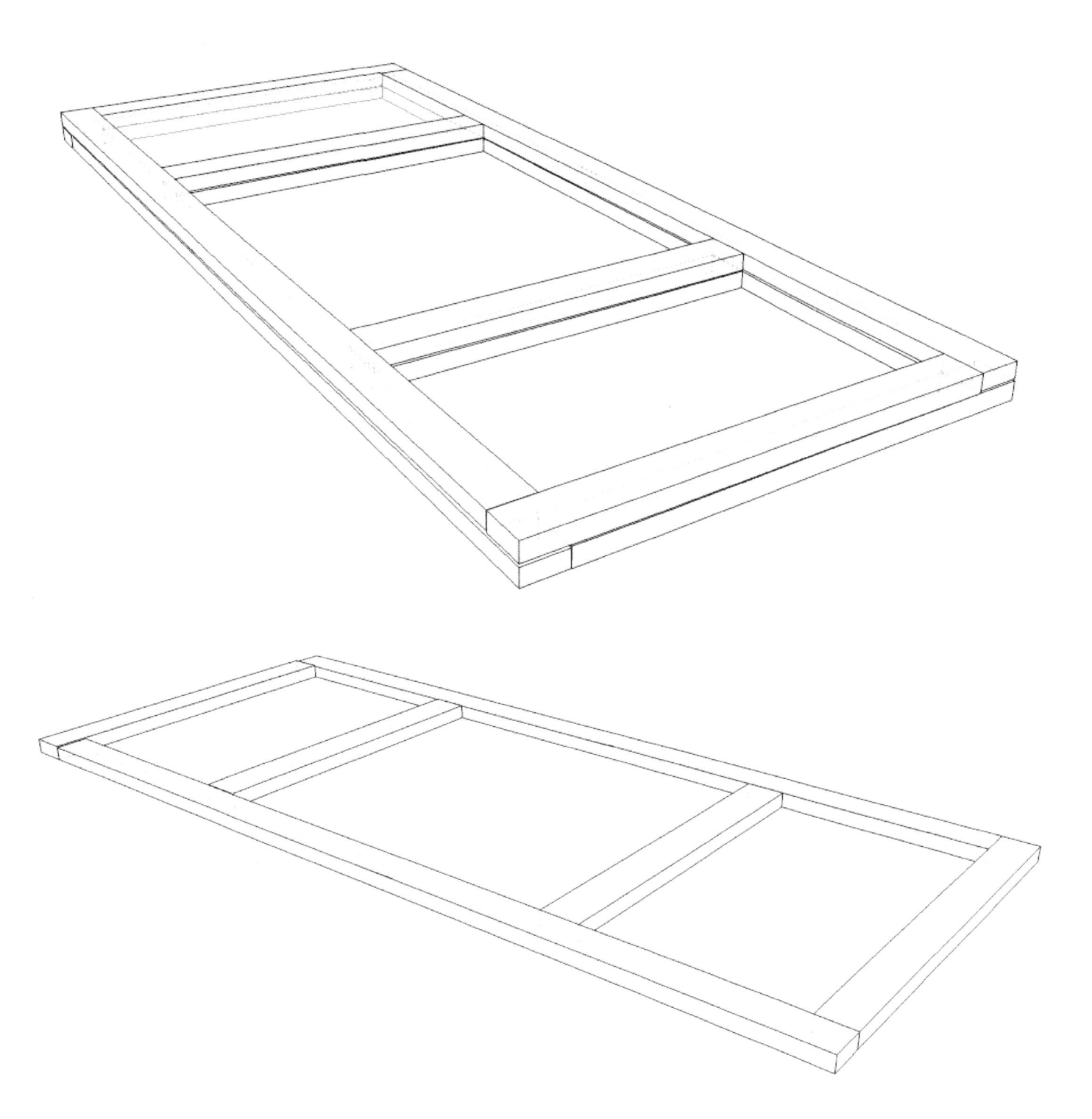

The hatchway door is fabricated from regular construction grade (or optional pressure treated) 2x4 lumber. The hatch can be made from a single layer of 2x4's as seen in the bottom illustration, and fastened together using staples and brackets, or, a two layer sandwich of 2x4's can be used as seen in the top illustration. Pressure treated lumber is advised for extended lifespan as the close proximity of the hatch to the ground in some designs may allow moisture contact. Use only screws and fasteners rated for use with pressure treated lumber.

This frame is covered with 1/2 inch plywood, then the entire exposed side is covered with a protective layer of 26 gauge galvanized sheet metal. This sheet metal layer must be cut to size, folded under on each side, and screwed or nailed down. Use of a sheet metal brake is recommended to do this properly. You can have the cover fabricated by a sheet metal shop if you do not have access to a brake, or cover it with common asphalt shingles.

H-12-4 block built shelter 3D floor plan.

H-12-4 block built shelter top view slab roof and hatchway opening.

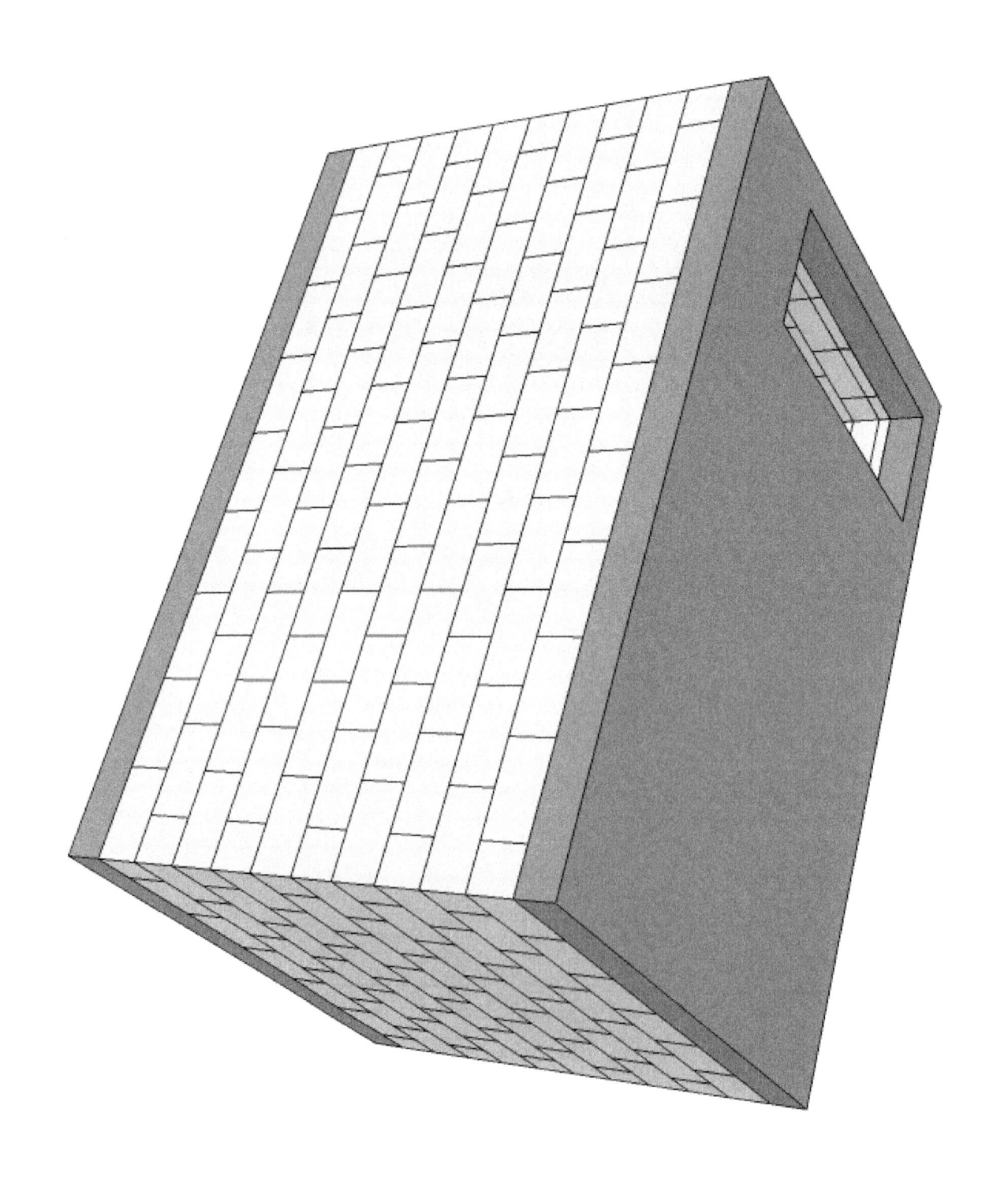

H-12-4 CMU (concrete masonry unit) walls with concrete slab foundation and roof with hatchway passage.

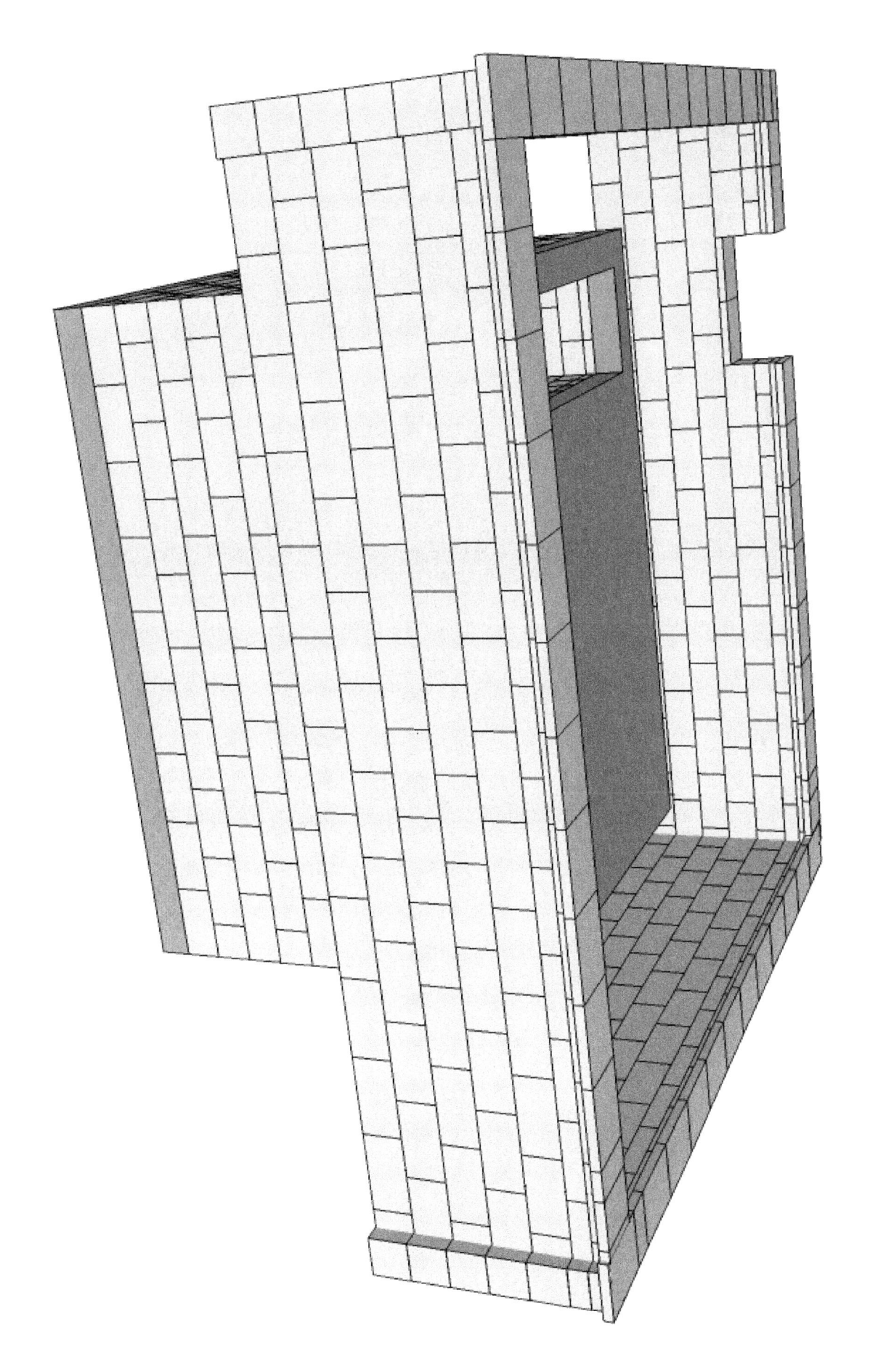

H-12-4 block built planter retaining wall surrounds the perimeter of shelter 2'-8" away on all sides.

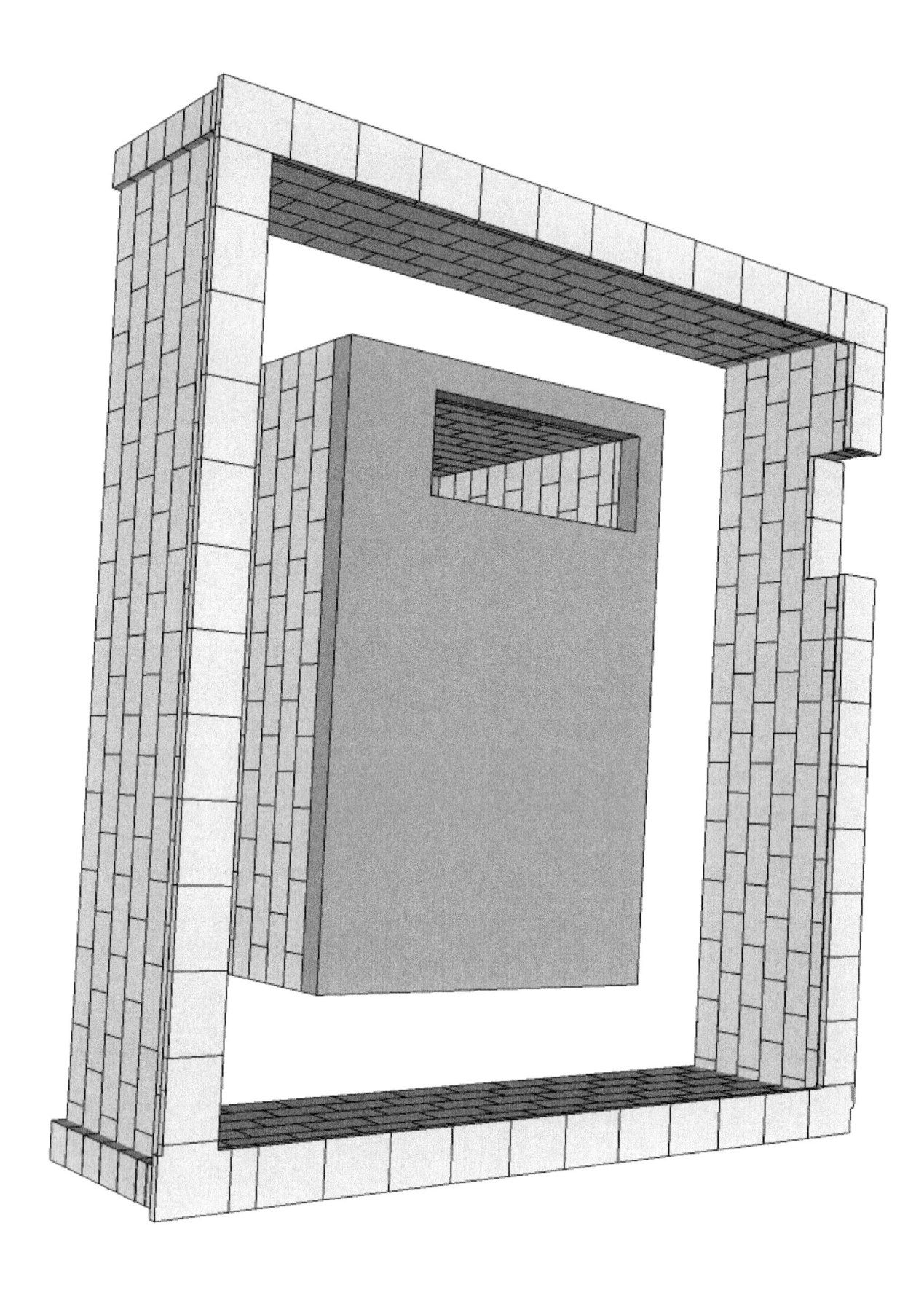

H-12-4 planter retaining wall surrounds the shelter and is filled with soil providing fallout protection inside the shelter with a shallower excavation.

H-12-4 the topside of planter retaining wall area can be landscaped (or paved with stone) in a variety of attractive ways for multiple use.

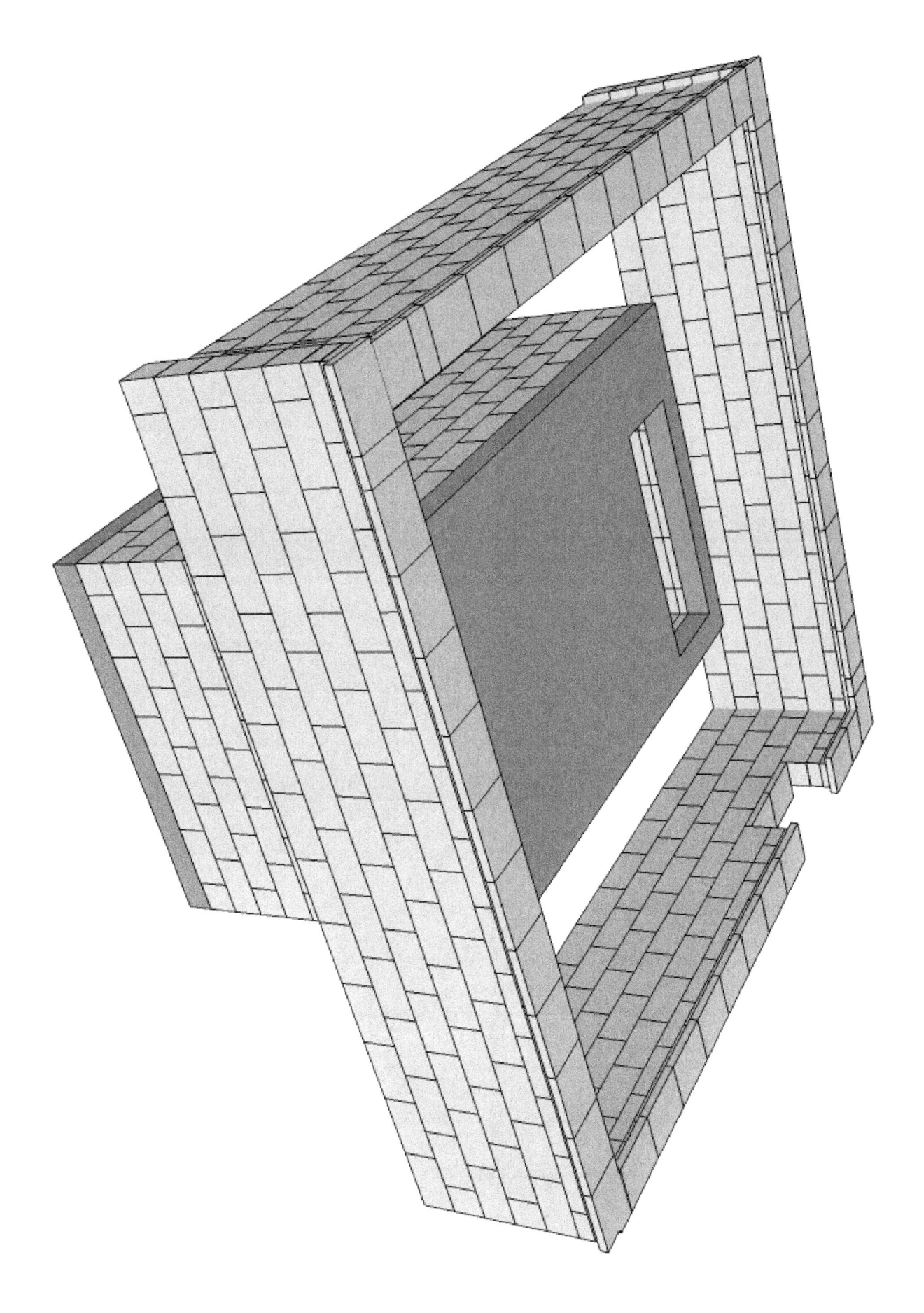

The H-12-4 plan utilizes ten rows of standard 8" CMU concrete masonry units for the underground shelter.

The standard 8" wide CMU block has the nominal dimensions of: 8" wide x 8" high x 16" long. The actual dimensions of concrete masonry block is 3/8" less (than what is listed above) to allow for mortar joints. These nominal dimensions work within the 4" grid that other construction materials follow.

Mortar is used to bond CMU blocks together. Choose type S, or stronger type M mortar for below grade applications.

Core-fill grout is thinner than mortar. It's a pour-able mixture of portland cement, small aggregate and sand that is pumped, poured, or dispensed using a grout bag to fill the CMU core cells.

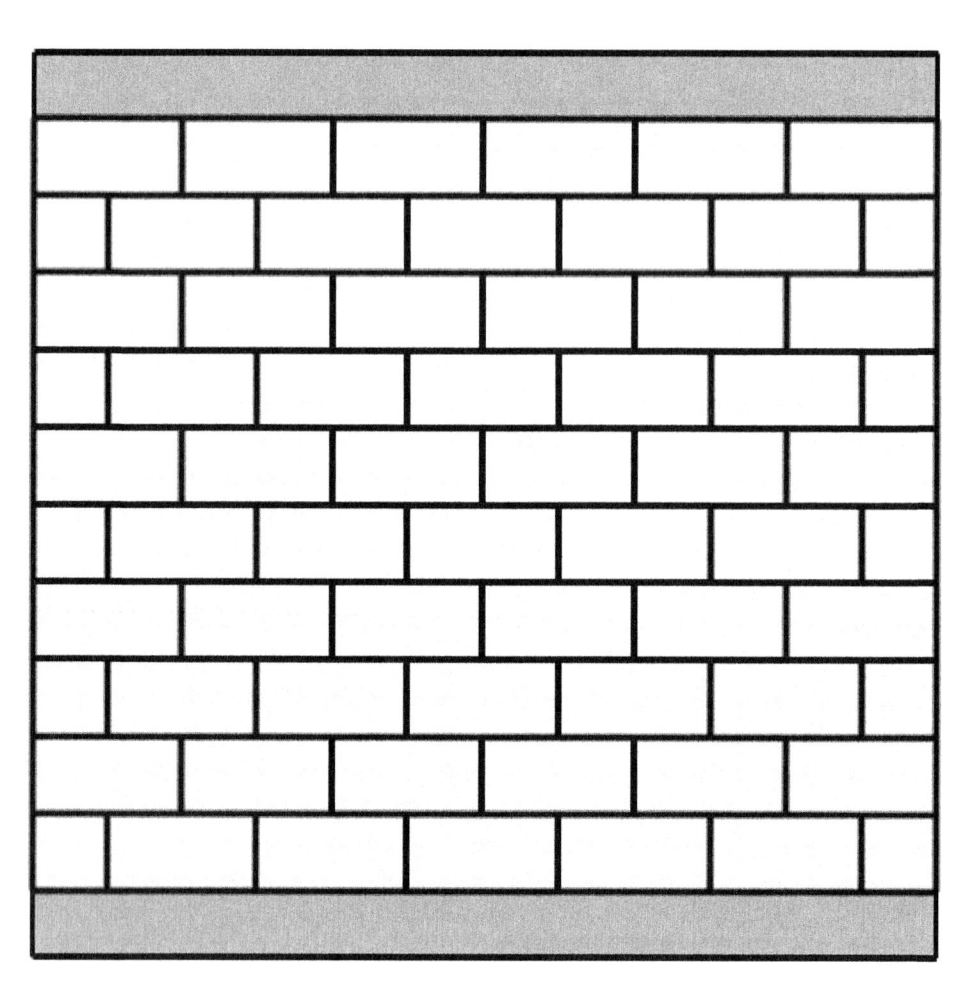

Notes

The form, or table, built for support during the pouring of the slab roof must be designed to hold thousands of pounds of wet concrete. It is advised that you "over build" rather than under build it. Under designed and built forms will result in a collapse during the pour.

I recommend obtaining a copy of 'Concrete, Masonry and Brickwork: A Practical Handbook for the Homeowner and Small Builder' published by the U.S. Army, prior to designing forms. This manual is used by the U.S. Army Corps of Engineers and covers all aspects of concrete and masonry work in great detail.

Form plywood must be thick enough (usually 1"), and support stud placement close enough, to prevent any warping or sagging during a pour. Form design must be such that form lumber can be removed by way of pre-planned disassembly after the concrete sets. The roof slab support form will need to be disassembled and removed through the small hatchway opening of the shelter roof after the concrete dries.

Before concrete is placed, forms are treated with oil or other coating material to prevent the concrete from sticking. The oil should penetrate the wood to prevent water absorption. A light bodied petroleum oil will do. On plywood, shellac or polyurethane is more effective than oil.

Wood wedges should be used to remove forms stuck against concrete rather than a metal pry bar that can damage uncured concrete.

Foam insulation panels can be cut to shape for use as spacers around hatchway (strips), and sump (disc), and so forth, to create recesses in the concrete where needed during a pour.

Slab floor can be finished to slope slightly toward sump pit for added drainage protection (optional).

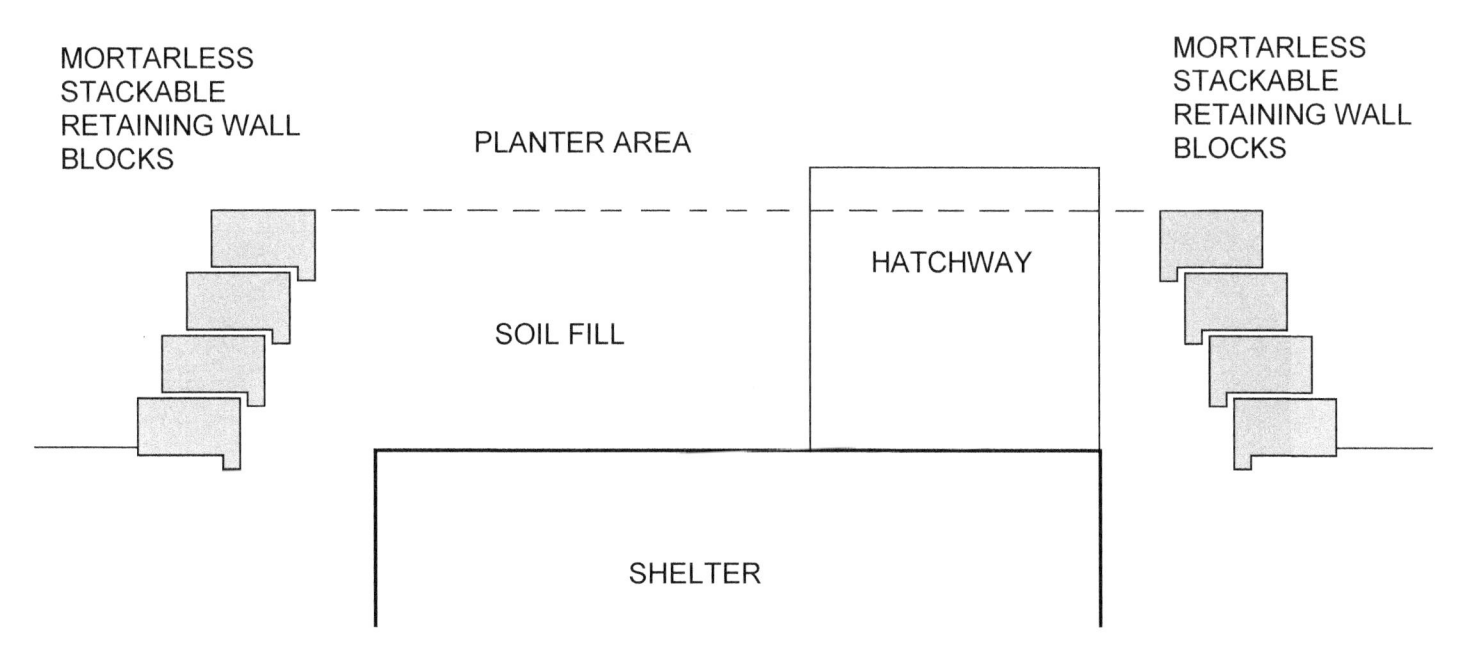

MORTARLESS STACKABLE RETAINING WALL BLOCKS

PLANTER AREA

MORTARLESS STACKABLE RETAINING WALL BLOCKS

HATCHWAY

SOIL FILL

SHELTER

Self-aligning, automatic setback retaining wall blocks are now commonly available at landscaping supply and home improvement stores. This type of block can be used to quickly erect an attractive retaining wall and may be considered as an alternative for building the CMU perimeter planter wall shown in shelter plans H-12-4.

Sump

The H-12-4 shelter plans call for installing a sump (MP-15 does not show a sump - but you should install one nevertheless). This is cheap insurance against water infiltrating the shelter, even if you are in a dry climate above the water table. Sumps are a must-have in area's where a shelter sits below a seasonably high water table or where seasonal heavy rain occurs.

Many people have heard of the term "sump" or "sump pump" but have a only rudimentary knowledge of what it is. Essentially a sump is a low space (most commonly a 18" diameter x 30" deep pit with a 2-3" bed of 1/2" gravel at the bottom) that collects excessive water from in and under the slab foundation of residential basements.

Larger diameter sump pits can be installed in high water table builds or areas where seasonal heavy rains are common. A larger sump pit diameter will hold more water resulting in less run time and longer pump life. Increasing the pit depth to 36" is also advised in these heavy duty applications.

In new construction projects (such as building a bomb shelter), the concrete floor slab is poured around the prefabricated plastic sump basin that has been placed (top level with the new floor) into a dug hole. Gravel is added to the bottom of the hole to bring the liner level with the top of its rim at floor level. The liner is then packed in place by pouring more 1/2" gravel around the sides and eventually it is cemented into a permanent position in the floor.

Optional under-slab perimeter perforated drain pipes (French drains) can also be installed before the pour that drain into the sump pit providing a higher level of dryness. This type of extra drainage is well worth the small investment of labor and material for the drying effect it provides inside the shelter.

The pumps used in sump pits are float switch operated, turning on-and-off as controlled by the water level inside the pit. You can use either a pedestal pump or a submersible pump. Pedestal pumps protrude from the pit offering a less than clean look, while submersible pumps are more susceptible to clogging problems... but remember, both types are proven designs and have been in use for around the world for decades. Submersible pumps have larger capacity and are less noisy to operate than pedestal pumps because they are installed down in the pit under the lid. Sump pumps are commonly 1/3 or 1/2 horsepower (200 or 400 W), either battery or electrically powered (or both). Pumps with a cast iron housing will disperse heat better and provide a longer service life than plastic models.

My personal recommendation for off-the-grid application is a 1/3 to 1/2 horsepower cast iron submersible sump pump with a deep cycle battery (solar powered main or backup) system.

Collected water is normally pumped out a 1-1/4" pipe with a check valve installed to prevent backflow. Accommodations to run this pipe through the slab roof should be made prior to the roof pour (1-1/2" round void located in the roof over the sump area - seal remaining gap with caulk). Water is pumped up from the sump and discharged to a designated drainage area at ground level away from the shelter. Sump pits are topped with a lid that fits around pipes and pedestal bases.

Sump pumps with 12V deep cycle battery powered backup systems are commonly available and can be used with a solar or wind powered charging system in remote areas. Use of a power inverter will allow common (non-backup) sump pumps to be powered by deep cycle batteries as well. The obvious drawback to these alternative energy battery charged designs is loss of pumping ability in the event of depleted batteries. Installing a manual hand operated bilge pump on the wall above the sump (available from boating supply stores) is recommended as a backup for off-the-grid, battery powered shelter setups.

Dehumidifiers can also be set-up to automatically drain into the sump.

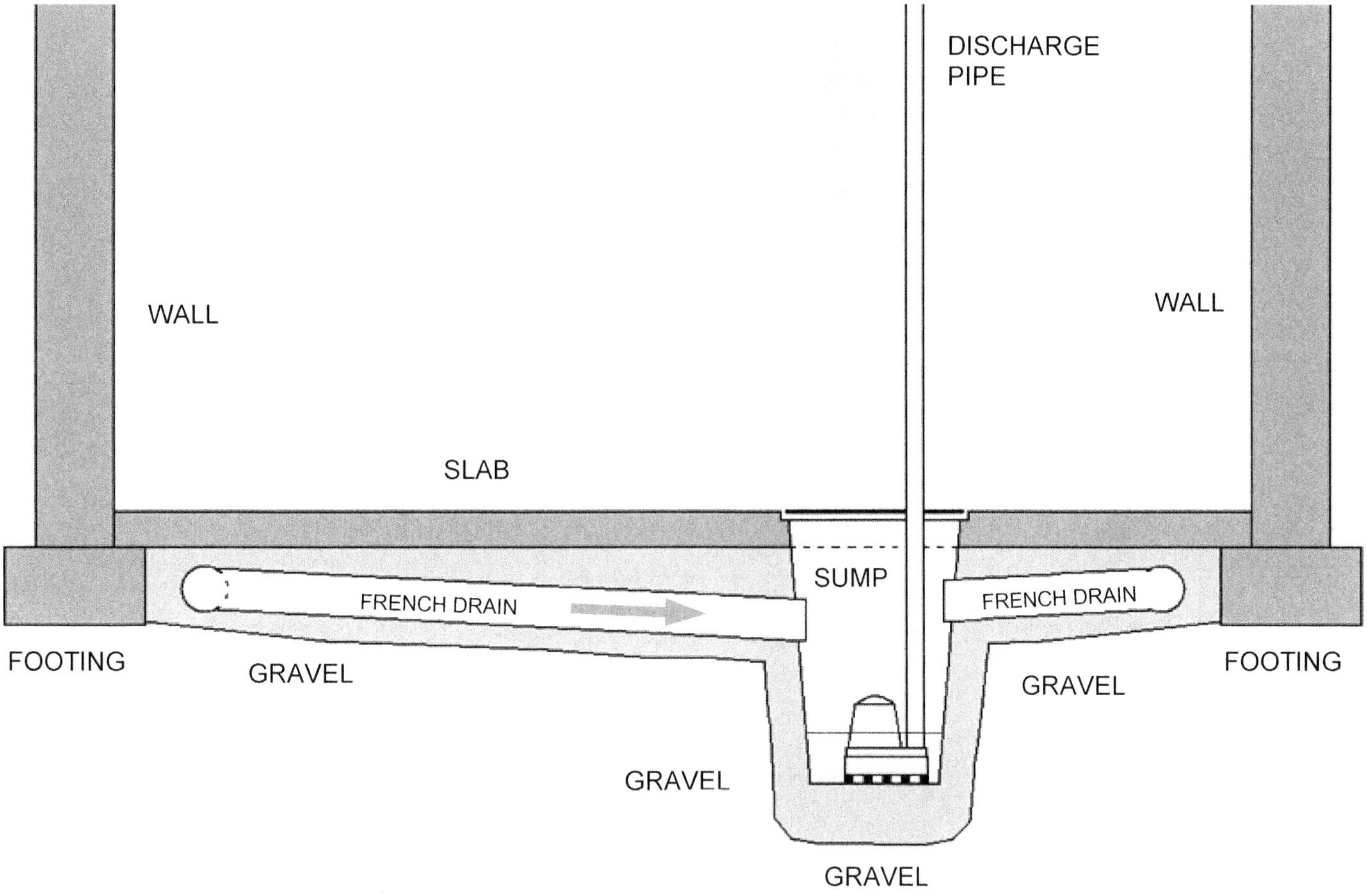
DISCHARGE
PIPE
WALL
WALL
SLAB
SUMP
FRENCH DRAIN
FRENCH DRAIN
FOOTING
GRAVEL
GRAVEL
FOOTING
GRAVEL
GRAVEL

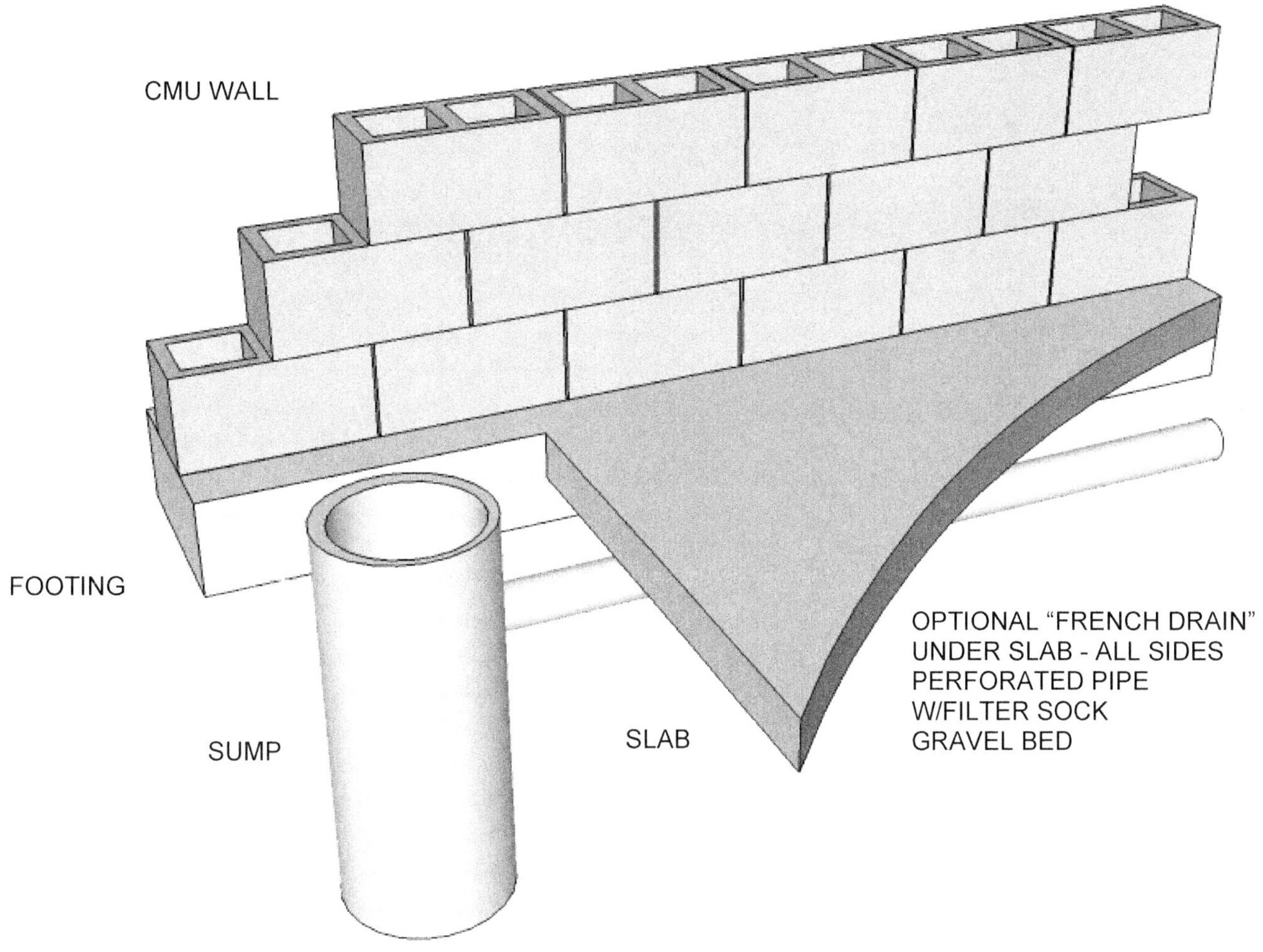
CMU WALL
FOOTING
OPTIONAL "FRENCH DRAIN"
UNDER SLAB - ALL SIDES
PERFORATED PIPE
W/FILTER SOCK
GRAVEL BED
SUMP
SLAB

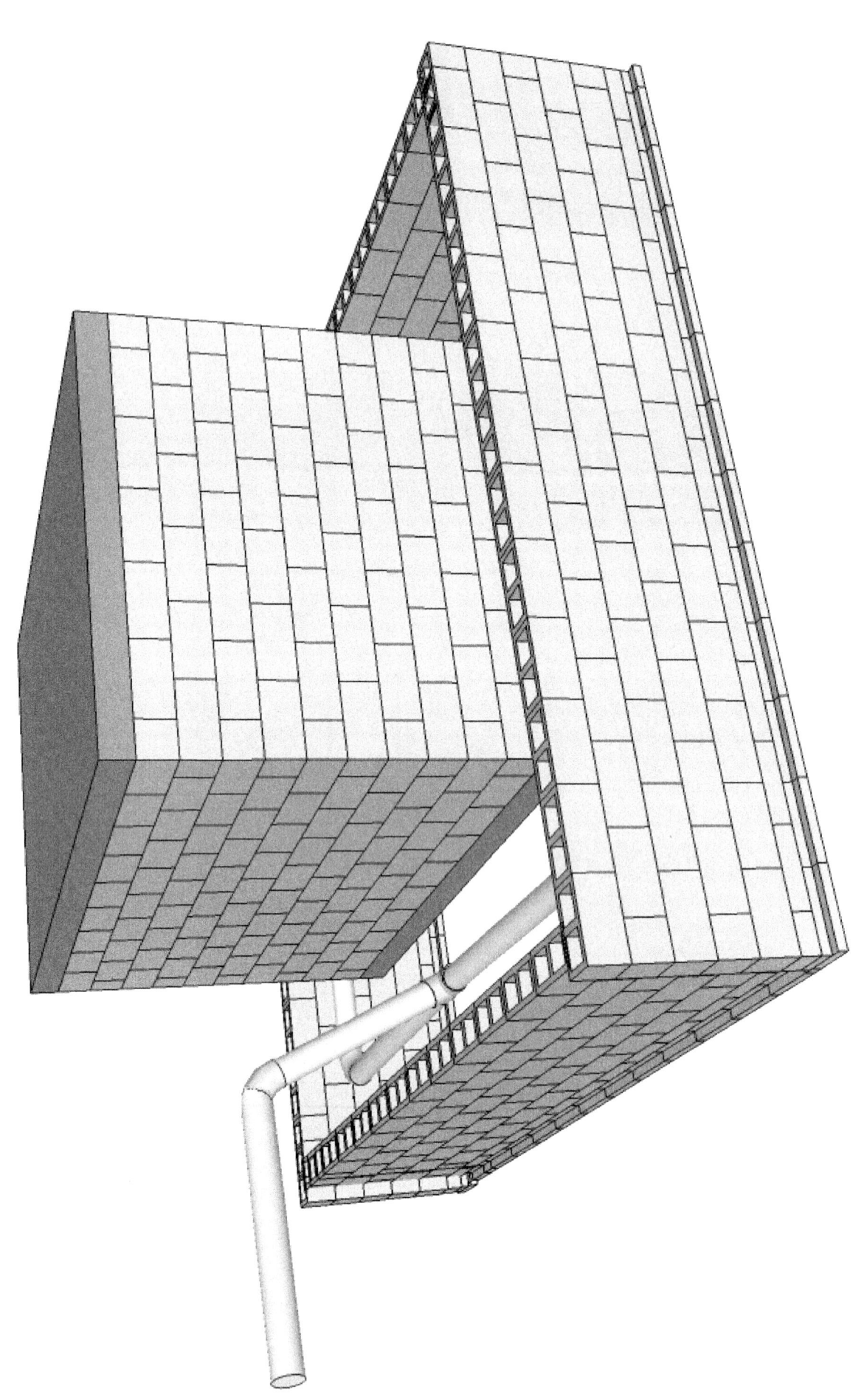

Shelters built on high ground have the benefit of gravity drainage that allows installation of a perimeter drainage system in the fill-area between the wall and shelter. A similar perimeter drain system can be installed around the shelter footer if site elevation permits drainage. Where site elevation does not permit running gravity fed drainage to a remote drain field, termination in a dry well is an acceptable option.

Foundation Drainage and Weatherproofing

Avoid the risk of developing leaks inside the shelter (due to pressure on the shelters walls and foundation created by high water tables or saturated ground conditions during periods of heavy rain) by installing 4" perforated drain pipe in a fabric filter sock, and cover with a 12" bed of gravel. Cover gravel bed with landscape fabric to keep sediment out. Drain dependant on availability of an outlet as discussed on the previous page.

If you install a modern drainage system like the one discussed here, you can skip wet-proofing the exterior walls and roof with liquid asphalt (as discussed on page 13) and install optional polystyrene foam insulation panels... but do not do both. Liquid asphalt (roofing type) compounds are petroleum-based and will dissolve polystyrene foam, so choose one or the other, but again, do not mix the two materials.

Concrete bonds exceptionally well with polystyrene, meaning you can place panels directly in forms, or apply panels after the pour. 2" thick insulation panels are used on the roof, and 1" thick insulation panels on exterior walls. Backfill the soil around the outside of the shelter walls and roof in 12" thick layers. Be careful not to push large rocks against the concrete or insulation.

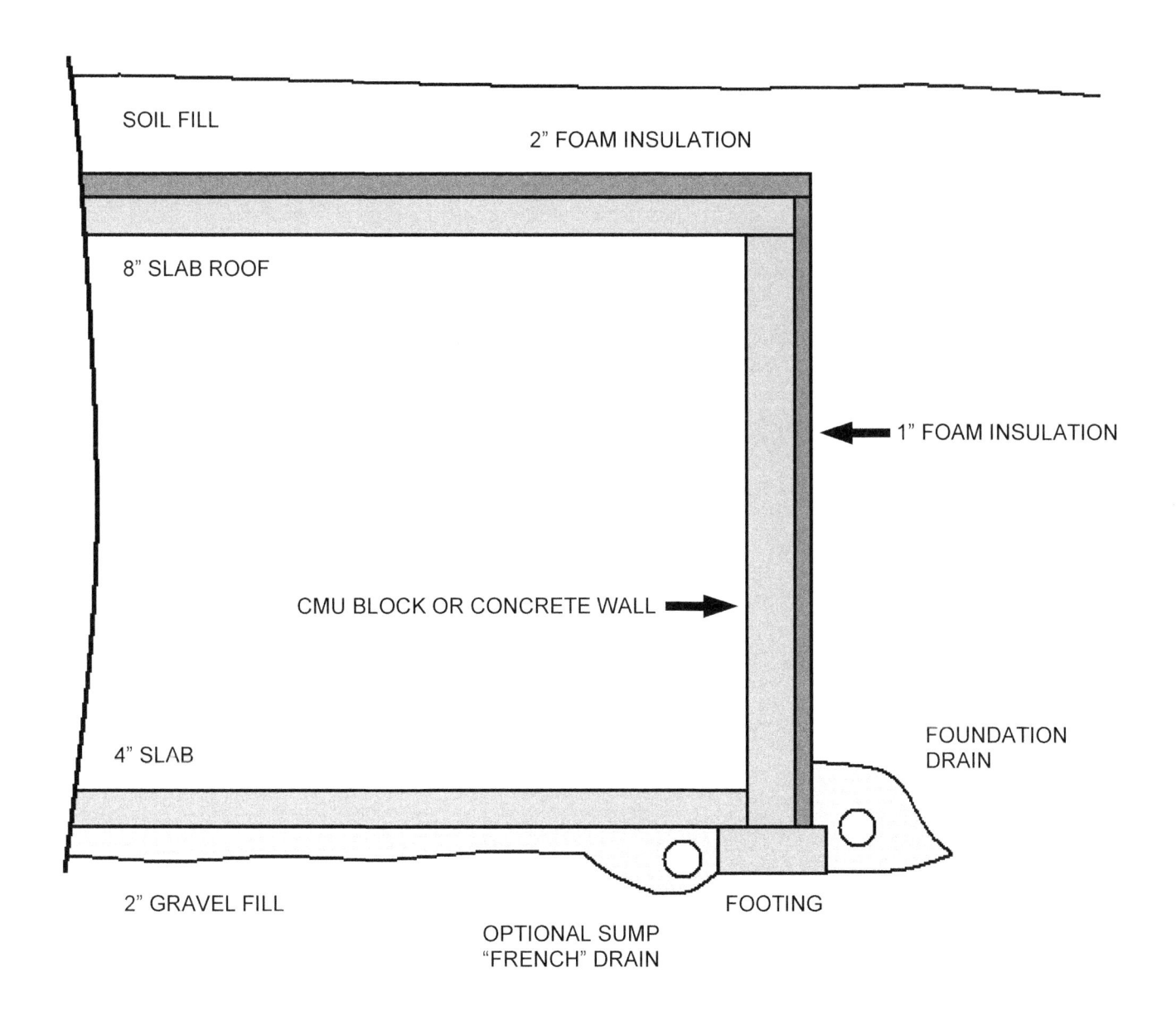

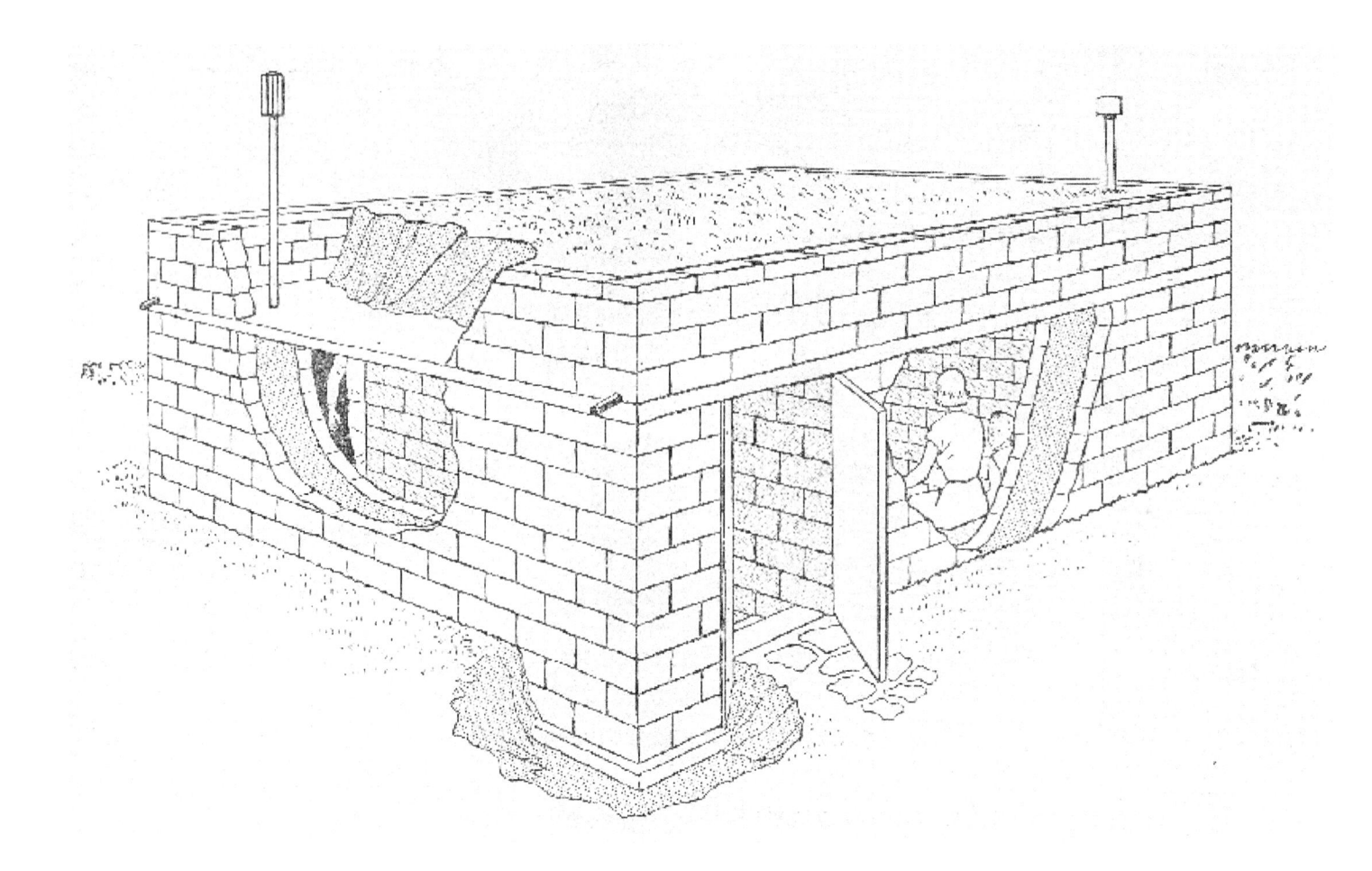

Aboveground Masonry Block Bomb Shelter

Booklet MP-15 The Family Fallout Shelter, also offered plans for building an aboveground double-wall masonry block shelter for areas where water tables or rock formations are close to the surface, making it impractical to excavate for an underground shelter.

Two walls of concrete blocks are constructed at least 20" apart. The space between them is filled with pit-run gravel or soil fill. The walls are held together with metal ties placed in the wet mortar as the walls are built. The roof is a 6" slab of reinforced concrete, covered with at least 20" of pit-run gravel. An alternate roof, perhaps more within do-it-yourselfers reach, could be constructed of heavy pressure treated beams, overlaid with 1" thick marine grade plywood and waterproofing. It would have to be covered with at least 28 inches of pit-run gravel for adequate fallout protection.

This design features double core-filled walls, but can also be modified with additional external earth sheltering.

Earth berming is a technique where soil backfill is piled up against exterior walls and packed, sloping down away from the shelter. There are fewer moisture control variables with earth berming in comparison to underground shelter construction.

In-hill construction is a method where the shelter is set into a slope or hillside. Three walls are embedded within the hillside with the entrance wall remaining open.

The average temperature of earth below the frost line is 55-57 degrees Fahrenheit. This means that any shelter design, whether it is underground, partially underground, or earth sheltered... will experience some benefit of a fairly constant temperature, even during the seasonal extremes of summertime heat and wintertime cold.

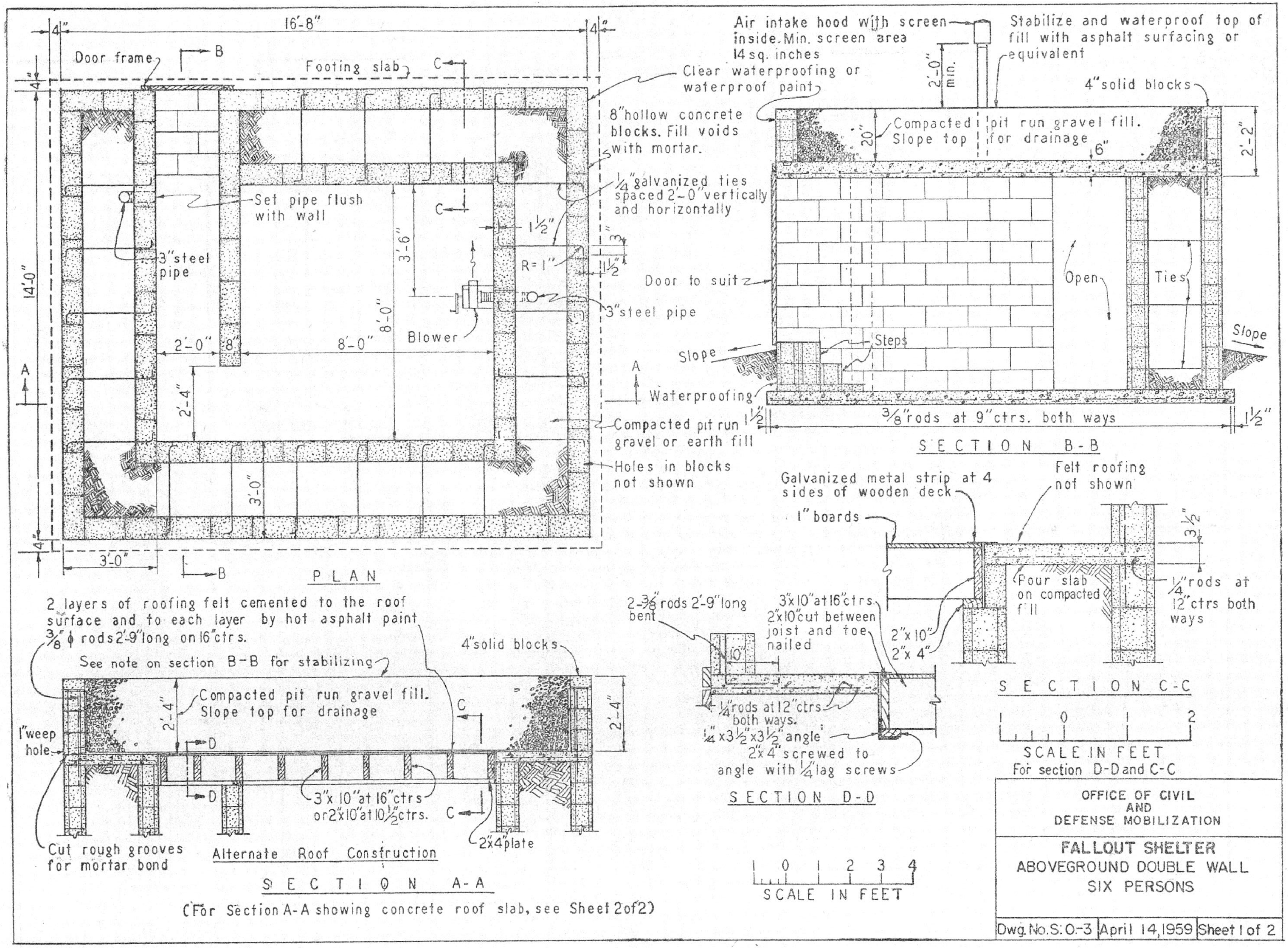
PLAN
Door frame
Footing slab
16'-8"
14'-0"
4"
3'-0"
2'-0"
8"
8'-0"
3'-6"
2'-4"
1½"
3"
R=1"
Blower
Set pipe flush with wall
3" steel pipe
Clear waterproofing or waterproof paint
8" hollow concrete blocks. Fill voids with mortar.
¼" galvanized ties spaced 2'-0" vertically and horizontally
Door to suit
Waterproofing
Compacted pit run gravel or earth fill
Holes in blocks not shown
2 layers of roofing felt cemented to the roof surface and to each layer by hot asphalt paint
⅜" ϕ rods 2'-9" long on 16" ctrs.
See note on section B-B for stabilizing
4" solid blocks
Compacted pit run gravel fill. Slope top for drainage
1" weep hole
Cut rough grooves for mortar bond
3"x10" at 16" ctrs or 2"x10" at 10½" ctrs.
2"x4" plate
Alternate Roof Construction
SECTION A-A
(For Section A-A showing concrete roof slab, see Sheet 2 of 2)
Air intake hood with screen inside. Min. screen area 14 sq. inches
Stabilize and waterproof top of fill with asphalt surfacing or equivalent
2'-0" min.
20"
6"
2'-2"
Compacted Slope top
pit run gravel fill. for drainage
Open
Ties
Steps
Slope
⅜" rods at 9" ctrs. both ways
SECTION B-B
Galvanized metal strip at 4 sides of wooden deck
Felt roofing not shown
1" boards
2-⅜" rods 2'-9" long bent
3"x10" at 16" ctrs. 2"x10" cut between joist and toe nailed
2"x10"
2"x4"
10"
¼" rods at 12" ctrs both ways.
¼"x3½"x3½" angle
2"x4" screwed to angle with ¼" lag screws
SECTION D-D
SCALE IN FEET
Pour slab on compacted fill
3½"
¼" rods at 12" ctrs both ways
SECTION C-C
SCALE IN FEET
For section D-D and C-C
OFFICE OF CIVIL AND DEFENSE MOBILIZATION
FALLOUT SHELTER
ABOVEGROUND DOUBLE WALL
SIX PERSONS
Dwg. No. S.O-3
April 14, 1959
Sheet 1 of 2

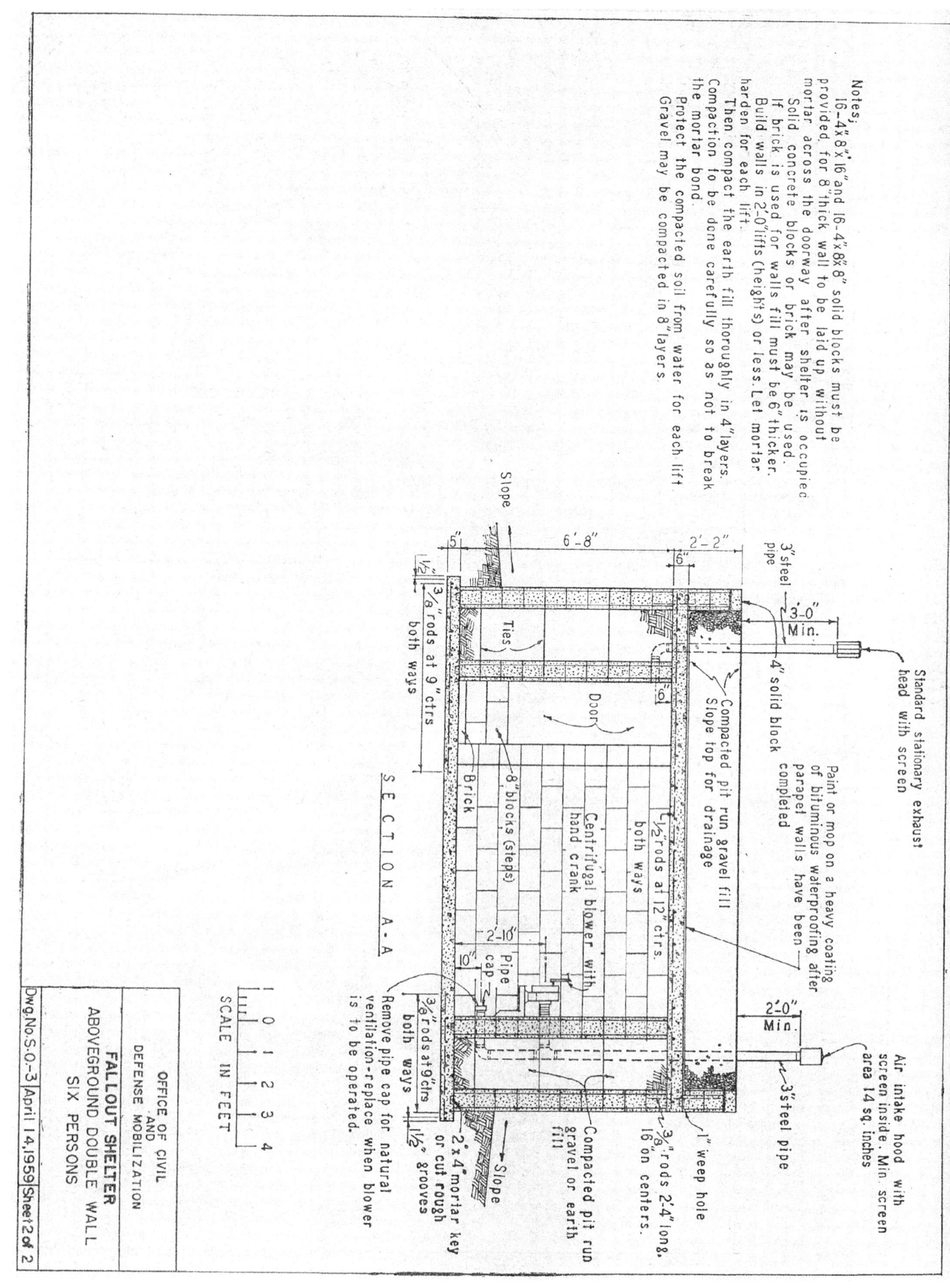

Notes:
16-4"x8"x16" and 16-4"x8"x8" solid blocks must be provided for 8" thick wall to be laid up without mortar across the doorway after shelter is occupied.
Solid concrete blocks or brick may be used.
If brick is used for walls fill must be 6" thicker.
Build walls in 2'-0" lifts (heights) or less. Let mortar harden for each lift.
Then compact the earth fill thoroughly in 4" layers. Compaction to be done carefully so as not to break the mortar bond.
Protect the compacted soil from water for each lift. Gravel may be compacted in 8" layers.

SECTION A-A

SCALE IN FEET: 1 0 1 2 3 4

OFFICE OF CIVIL AND DEFENSE MOBILIZATION
FALLOUT SHELTER
ABOVEGROUND DOUBLE WALL
SIX PERSONS

Dwg.No.S.O-3 | April 14,1959 | Sheet 2 of 2

Labels on drawing:
- Standard stationary exhaust head with screen
- 3" steel pipe
- Ties
- Compacted pit run gravel fill. Slope top for drainage completed
- 4" solid block
- Paint or mop on a heavy coating of bituminous waterproofing after parapel walls have been completed
- Door
- Brick
- 8" blocks (steps)
- Centrifugal blower with hand crank
- Pipe cap
- 3/8" rods at 9" ctrs both ways
- 1/2" rods at 12" ctrs both ways
- 3/8" rods at 9" ctrs both ways
- 1/2" grooves
- 2x4" mortar key or cut rough
- Remove pipe cap for natural ventilation - replace when blower is to be operated
- Compacted pit run gravel or earth fill
- 1" weep hole
- 3" steel pipe 2'-4" long, 16 on centers
- Air intake hood with screen inside. Min. screen area 14 sq. inches
- Slope

PLAN

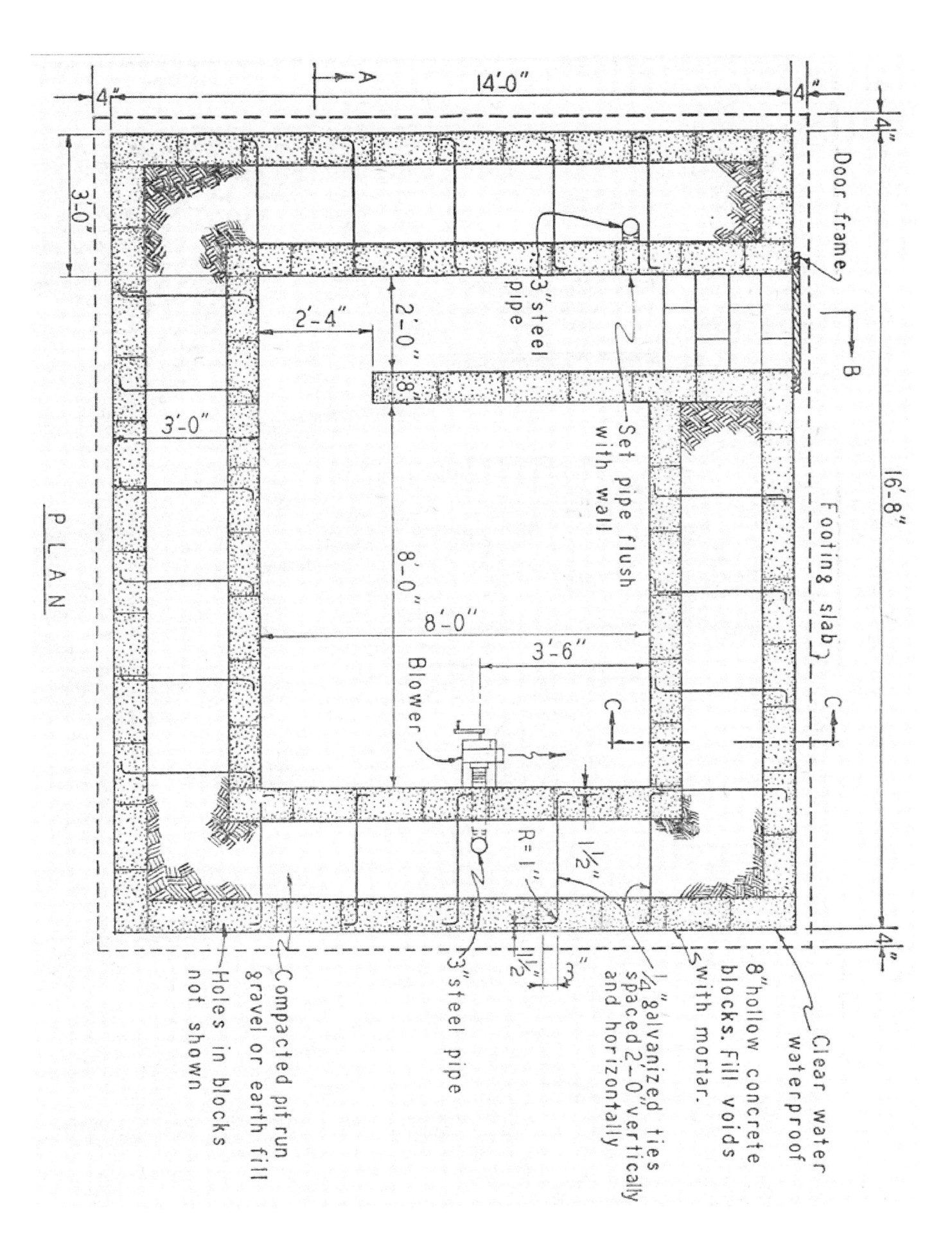

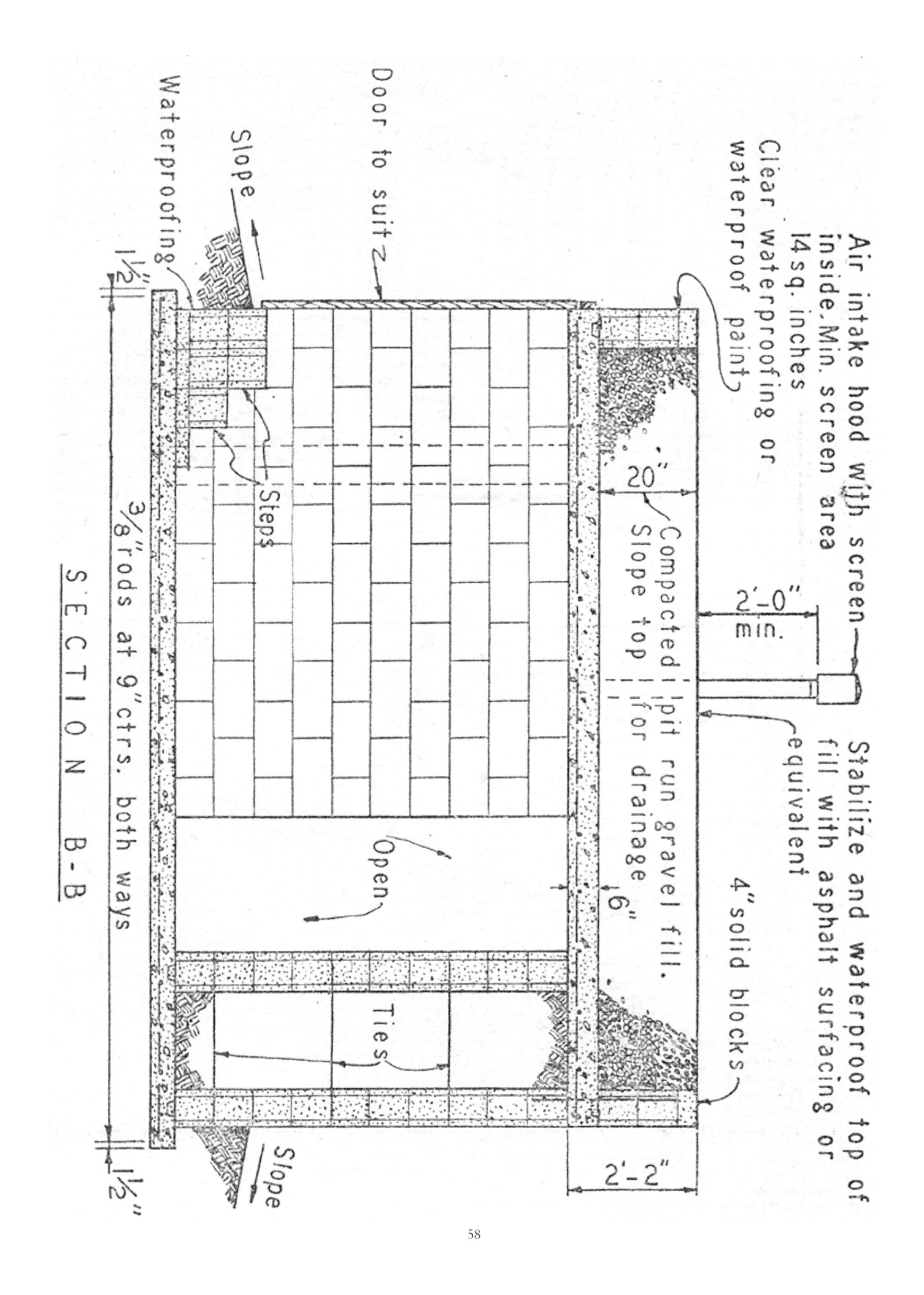

Air intake hood with screen inside. Min. screen area 14 sq. inches

Clear waterproofing or waterproof paint

Door to suit

Waterproofing

Slope

1½"

3/8" rods at 9" ctrs. both ways

Steps

20"

Compacted pit run gravel fill. Slope top for drainage

2'-0" min.

Stabilize and waterproof top of fill with asphalt surfacing or equivalent

4" solid blocks

Open

6"

Ties

Slope

1½"

2'-2"

SECTION B-B

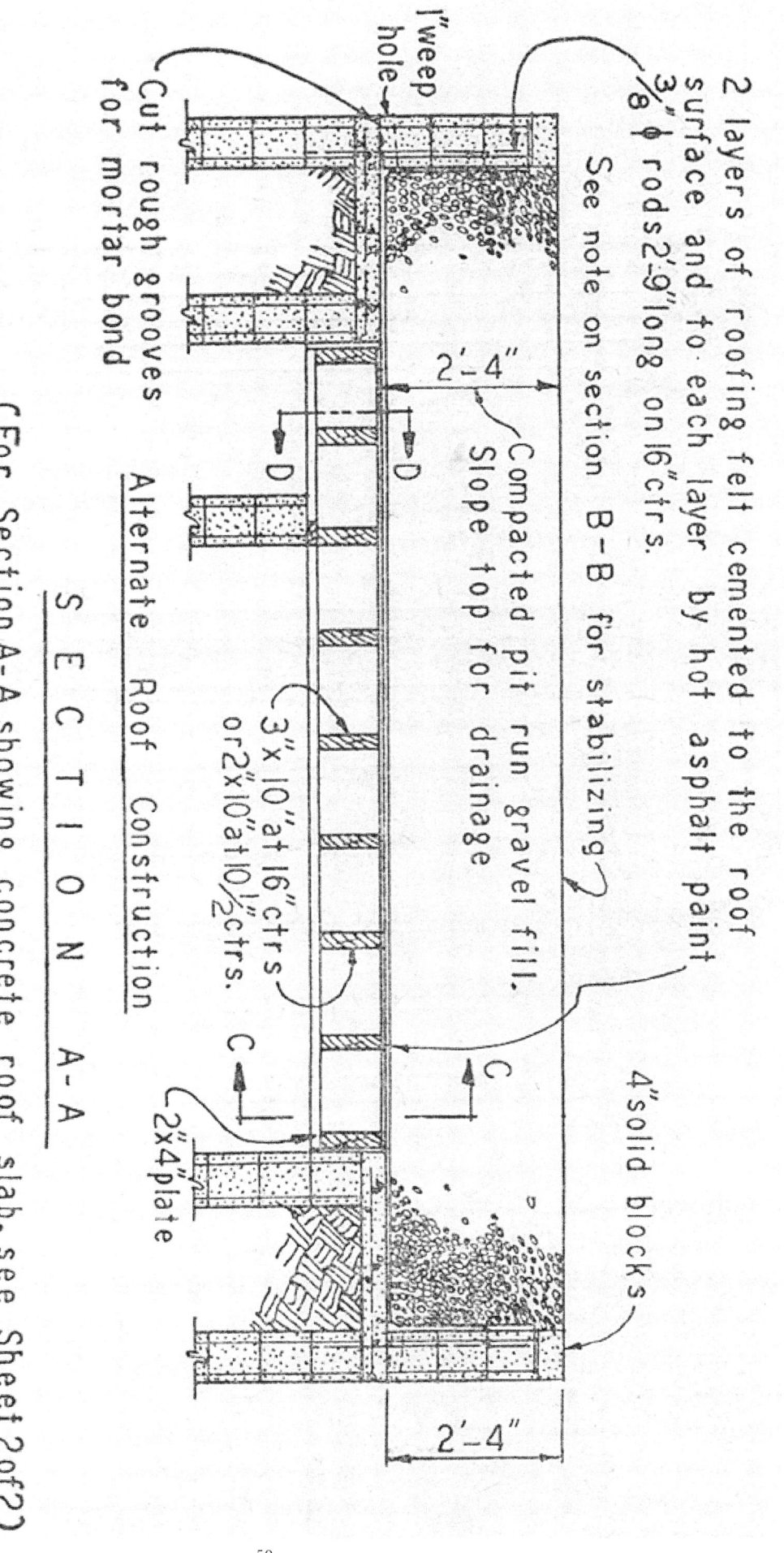

2 layers of roofing felt cemented to the roof surface and to each layer by hot asphalt paint

3/8"∅ rods 2'-9" long on 16" ctrs.

See note on section B-B for stabilizing.

1" weep hole

2'-4"

Compacted pit run gravel fill. Slope top for drainage

3"x10" at 16" ctrs or 2x10"at 10½"ctrs.

Cut rough grooves for mortar bond

2"x4" plate

4" solid blocks

2'-4"

SECTION A-A
Alternate Roof Construction
(For Section A-A showing concrete roof slab, see Sheet 2 of 2)

2-⅜" rods 2'-9" long bent

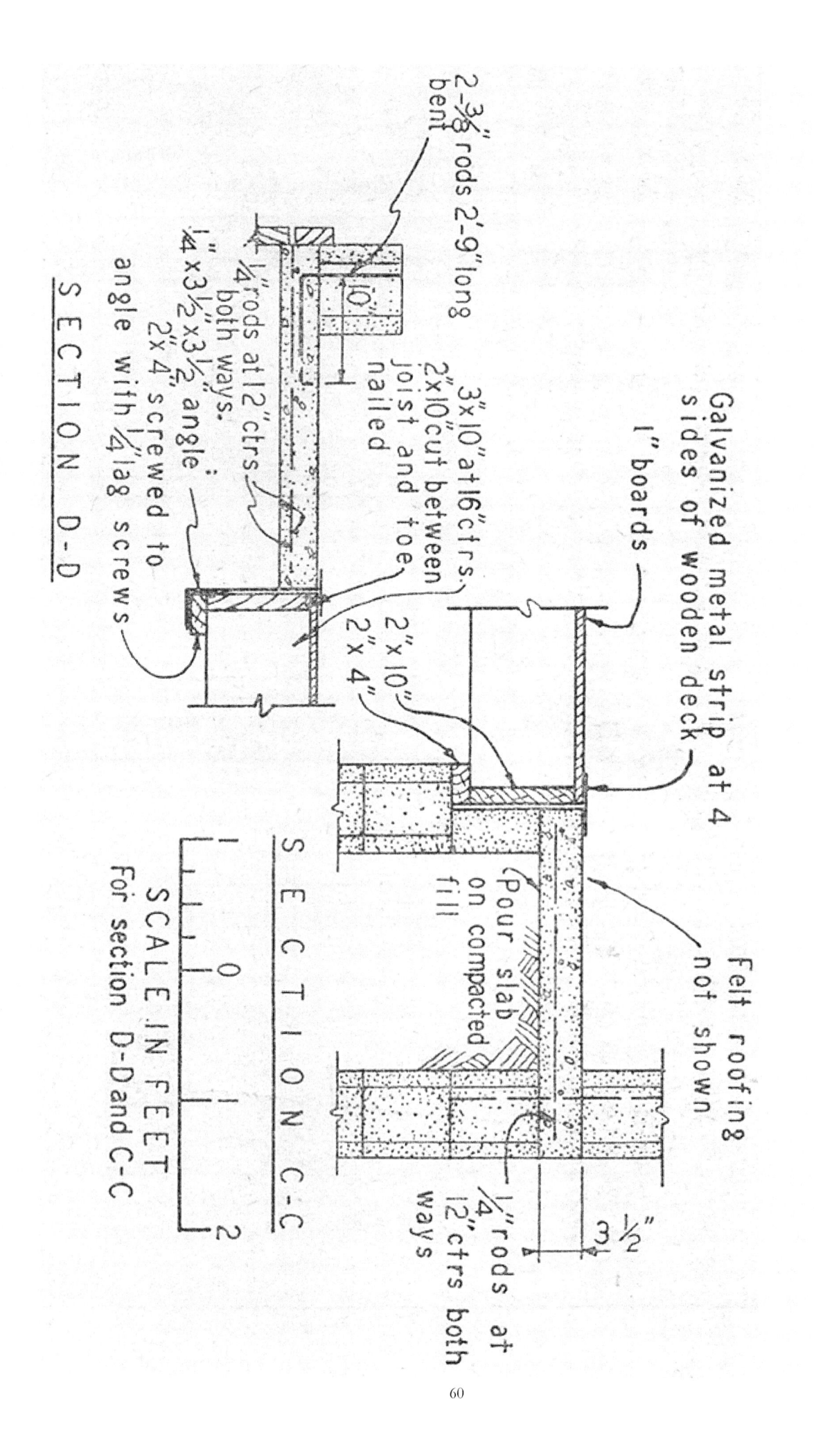

Galvanized metal strip at 4 sides of wooden deck

1" boards

¼" rods at 12" ctrs. both ways.

3"x10" at 16" ctrs. 2x10 cut between joist and toe nailed

¼" x 3½" x 3½" angle; 2"x 4½" screwed to angle with ¼" lag screws

2"x10"
2"x4"

Pour slab on compacted fill

Felt roofing not shown

¼" rods at 12" ctrs both ways

SECTION D-D

SECTION C-C

SCALE IN FEET
For section D-D and C-C

60

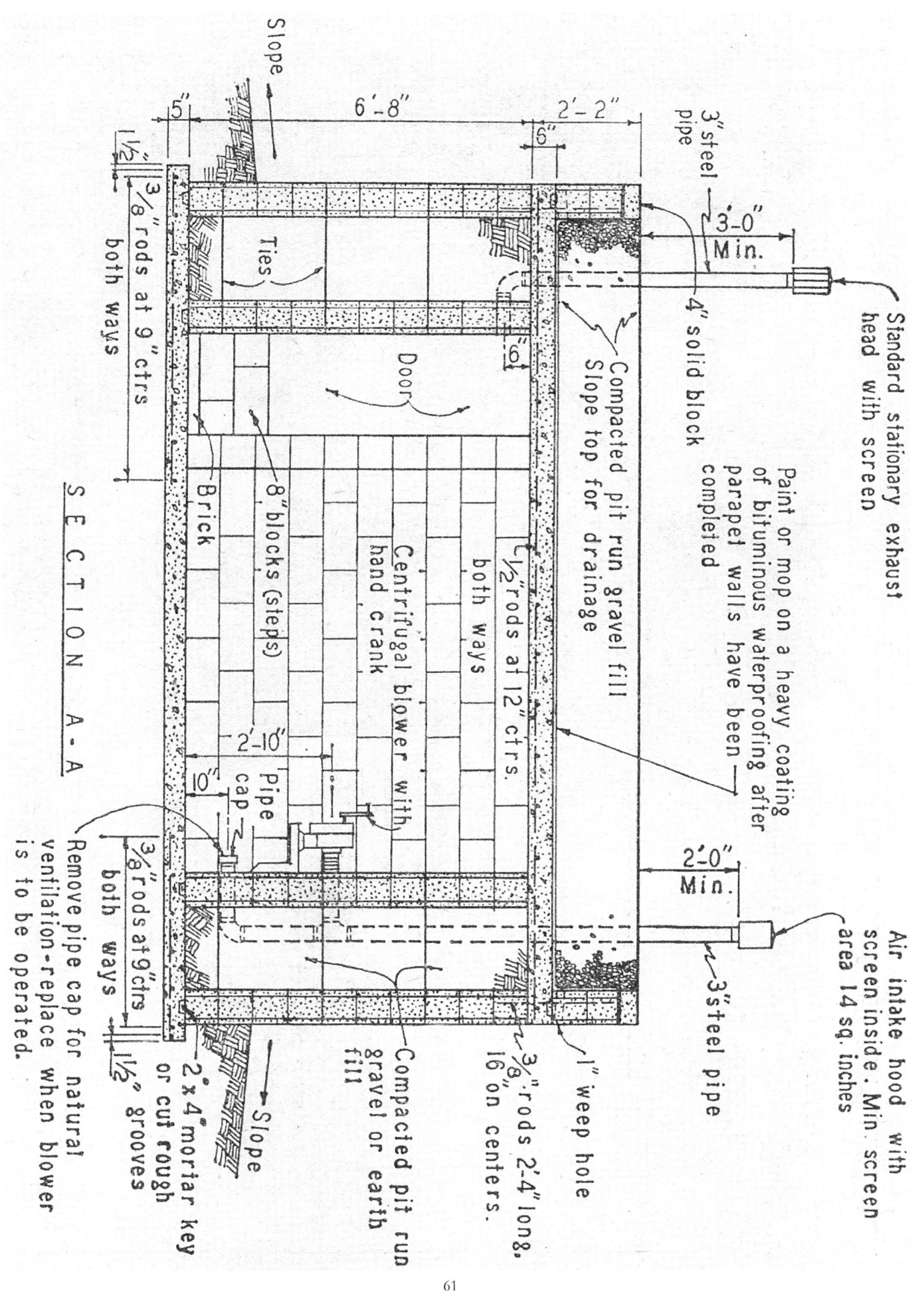

SECTION A-A

- 3" steel pipe
- 6'-8"
- 2'-2"
- 5"
- 6"
- 1½"
- ⅜" rods at 9" ctrs both ways
- Slope
- Ties
- 4" solid block
- 3'-0" Min.
- Standard stationary exhaust head with screen
- Compacted pit run gravel fill. Slope top for drainage
- Paint or mop on a heavy coating of bituminous waterproofing after parapet walls have been completed
- Door
- 6"
- Brick
- 8" blocks (steps)
- ½" rods at 12" ctrs. both ways
- 2'-10"
- 10"
- Pipe cap
- Centrifugal blower with hand crank
- ⅜" rods at 9" ctrs both ways
- 1½"
- 2 x 4 mortar key or cut rough
- 1½" grooves
- Slope
- Compacted pit run gravel or earth fill
- ⅜" rods 2'-4" long, 16" on centers.
- 1" weep hole
- 2'-0" Min.
- 3" steel pipe
- Air intake hood with screen inside. Min. screen area 14 sq. inches
- Remove pipe cap for natural ventilation - replace when blower is to be operated.

61

Printed in Great Britain
by Amazon